T0206950

Springer Biographies

The books published in the Springer Biographies tell of the life and work of scholars, innovators, and pioneers in all fields of learning and throughout the ages. Prominent scientists and philosophers will feature, but so too will lesser known personalities whose significant contributions deserve greater recognition and whose remarkable life stories will stir and motivate readers. Authored by historians and other academic writers, the volumes describe and analyse the main achievements of their subjects in manner accessible to nonspecialists, interweaving these with salient aspects of the protagonists' personal lives. Autobiographies and memoirs also fall into the scope of the series.

David M. Behrman · Edward J. Behrman

Emil Fischer's "From My Life"

English Translation of "Aus Meinem Leben"

 Springer

David M. Behrman
Department of Mathematics and Chemistry
Somerset Community College
Burnside, KY, USA

Edward J. Behrman
Department of Chemistry and Biochemistry
The Ohio State University
Columbus, OH, USA

ISSN 2365-0613   ISSN 2365-0621 (electronic)
Springer Biographies
ISBN 978-3-031-05158-6   ISBN 978-3-031-05156-2 (eBook)
https://doi.org/10.1007/978-3-031-05156-2

This Springer imprint is published by the registered company Springer Nature Switzerland AG
The registered company address is: Gewerbestrasse 11, 6330 Cham, Switzerland

From My Life, written in that year of misfortune, 1918

Preface

This book is a translation into English of Emil Fischer's autobiography first published in 1922, three years after the author's death. Emil Fischer was one of the greatest chemists of all time who carried out fundamentally important work on sugars, purines, and peptides. His proof of the stereochemistry of glucose remains one of the great intellectual achievements of science. And, it was largely for this, and for his work on purines, that he received the Nobel Prize in 1902. This translation will benefit current and future generations of chemists who, unlike past generations, are not proficient in German. It has been translated into Russian [1] but never before into English. The original edition [2] has been reprinted several times recently but without comments, additions, or illustrations. An exception is Witkop's 1987 reprint of the German text [3] which also contains a 24-page prologue and a 17-page epilogue, both in English. These should be read together with this translation; the prologue for its summary of Fischer's work, his thinking, and other matters, and the epilogue for its listing of Fischer's 310 students and co-workers compiled by Joseph Fruton. Fruton later expanded this survey [4]. Two splendid reviews of Fischer and his work were published in 2002, the centenary of his Nobel Prize. Lichtenthaler's [5], while dealing with many other matters, is particularly notable for its history of Fischer's scientific progeny while Kunz [6], after a biographical sketch, gives an excellent summary of Fischer's chemical discoveries.

Somerset, KY, USA
Columbus, OH, USA

David M. Behrman
Edward J. Behrman

References

1. Iz moeĭ Zhizni:avtobiografiya/Emil Fischer. Perevod s nemetskogo [translated from the German] by N. S. Gelman, Otvetstvennyi Redaktor [national editor] A. N. Shamin. Moskva, Nauka, 1988. This translation omits the three portraits of Fischer found in the original edition and in Witkop's edition but it contains a valuable name index as well as brief summary of Fischer's chemical work by the editor, Shamin. There is also a preface by Academician Yu. A. Ovchinnikov.

2. E. Fischer, Aus Meinem Leben, Verlag von Julius Springer, Berlin, 1922.
3. E. Fischer, Aus Meinem Leben, With a Prologue and Epilogue by Bernhard Witkop, Springer Verlag, Berlin, 1987.
4. J. S. Fruton, Contrasts in Scientific Style, Amer. Phil. Soc., Philadelphia, 1990, Appendix 6.
5. F. W. Lichtenthaler, Eur. J. Org. Chem. **2002**, 4095–4122.
6. H. Kunz, Angew. Chem. Int. Ed., **41**(2002)4439–4451.

Contents

Chapter 1
Youth

If you carry sorrow and the heavy burden of old age in your heart then call up the memories of your youth to visit you.
Kussmaul

I was born in the small regional city of Euskirchen, which lies on a fertile plain bordered on the south by the foothills of the Eifel and on the east by hills of the Ville. Euskirchen is near the Erft river, which rises in the Eifel and joins the Rhine at Neuss. The many brooks flowing through the city are very useful for the local cottage industries and gave us boys rich opportunities to practice our water skills. Sixty years ago, the city had perhaps 3500 inhabitants, among whom at least half made a living from farming the fertile surroundings. In addition, there was an old cloth industry which worked chiefly for the army. During the war years of 1864–1866 and 1870 numerous orders for standard military issue brought prosperity to the city. Somewhat less rewarding was the tanning industry, for which use was made of the abundant oak trees in the neighboring forests. Finally, the city was a center for small business for the region. These businesses held an advantage, since there was no competition within 30 km. Later, when the railroads came, the city became a hub, which lent further advantage to the development of industry. The first railroad was opened in 1862, an astonishing development for the entire population, not the least for the youth. The same was true at the introduction of gas lighting, which occurred about the same time.

In earlier centuries, the city had been fortified. In my youth, some parts of the old walls with towers, ramparts, and moats were still in place, providing us much entertainment in our leisure time. Particularly familiar to us was the so-called Jewish rampart, where we frequently hung around playing and rooting around the floor, and from where we cast astonished glances at the operation of a [nearby] tannery.

My parents' house lay on the road to Cologne, a few minutes from the old city. The building complex consisted of two spacious dwelling houses, one of which was occupied by my uncle, a [main] building for the business, together with various small workshops, as well as several modest farming buildings surrounding a large field. Gardens were bordered on one side by a brook, and a small ditch passed through.

© The Author(s), under exclusive license to Springer Nature Switzerland AG 2022
D. M. Behrman and E. J. Behrman, *Emil Fischer's "From My Life"*,
Springer Biographies, https://doi.org/10.1007/978-3-031-05156-2_1

I was born here on October 9, 1852 as the eighth and last child of my parents. Of my siblings, one boy and one girl had previously died. All the others were sisters, of whom the eldest was fourteen years my senior. One can well imagine, that under these circumstances my arrival was cause for great joy to my parents, and that later I would be given preference.

The sisters evinced interest in [their] sole brother, whom they called simply "the boy", in exceedingly many ways, and I had to resist these efforts, with the result that an aversion to young ladies persisted throughout my boyhood years. Luckily, the situation at my uncle's house was reversed, in that there were five sons and one daughter, who, oddly enough, was the youngest. Uncle's house was across the field, but one could also get there through the warehouse or via a special corridor between the houses. When the excessively feminine treatment in my house became too much, with my parents' permission I moved to the neighboring house for a few days. My stronger cousins, however, treated me with such blows that I [soon] returned to the gentler environment of my own house.

My father, Laurenz Fischer, together with his brother August, operated a commercial business, chiefly in groceries, wine and spirits. In addition, they owned a wool-spinning plant, which was located in the village of Wisskirchen, about an hour away from Euskirchen. The plant was originally powered by water but later by steam. Another uncle, Friedrich Arnold, was also involved in the entire business, [but] he lived in Flamersheim, the original location of the family, and administered the property that had originated with [my] grandfather.

In these portrayed surroundings of Euskirchen I spent an overall happy childhood. The commercial business, that supplied the shopkeepers in the surrounding area extending far into the Eifel, brought with it an industrious lifestyle. The traffic in the counting house, the warehouse, and over the field reminded me of the portrayal of the Breslau business in *Soll und Haben*, the novel by Gustav Freytag. Certainly, the relationships we had were unassuming, despite the success of the firm. All those [conducting business] were dressed in many merry colors. The vivacious style of the Rhenish population and the happy, good-humored disposition[s] of the shopkeepers provided for many lively and jesting conversations, in spite of the orderly business procedures where the many strict rules of respectability had to be followed.

A small agricultural business, that satisfied the household needs, provided in addition a manifold drama, interesting for us children. Imagine twelve young people forming a single family over field and garden. Later, when my sisters were married, the grandchildren increased our number. One may well imagine the multitude of leisure activities that we in this circle engaged in. In early childhood, our play assumed various forms, from ball games to Indian wigwams, catching fish and birds, camping, and then fighting in various ways among us boys or in closed ranks against evil forces. Battles involved not just fists and sticks, but also stone-throwing and the catapult, rising to life-threatening level, which then had to be stopped by adults.

We stood on trusted footing with the many servants, particularly the men, and the conversations were exclusively in the coarse lower-Rhine dialect. Naturally we all had nicknames. I was nicknamed "Baron", though whether this was due to a sumptuous diet or better clothing has remained a mystery to me.

There was never a question of coddling [the children], neither physically nor emotionally. In the lightest clothing we knocked each other around throughout the winter, and the only injuries I remember from early childhood came from panaritium of the fingers, from frozen feet, or from boots that had become hard and narrow from [exposure to] meltwater. In playing on the ice, I fell through many times, once even into a cesspool over my head. When I returned home in this condition, filthy and evil-smelling, I had to undress outside, despite the sharp cold. All of this caused no [lasting] damage.

Another time I fell from a wagon loaded with bales of wool onto the road and entered the house with a severely bleeding head wound. I was greeted with these cheerful words: "Better a hole in the head than one in the trousers."

Certainly, there were more severe accidents. When playing with powder, through reckless [behavior] of a comrade, I was severely burned on face and head. This caused great alarm, for as I was led with closed eyes by an old lady before my mother, she feared I had been blinded and broke out sobbing. Luckily, I came through this accident, again without lasting damage, a result of good medical care, and I had the satisfaction, through obstinate silence, of avoiding incriminating the other guilty parties. The feeling of solidarity among us boys was highly developed, particularly true in school, where every lie was justified in our eyes when it served to protect comrades from punishment.

It was customary at the time to send children to school at age five, and that same fate became my part; for one day after my fifth birthday, I was taken along to school by my older sisters.

The *Volksschule* in my hometown was under the influence of the Catholic Church, and genuine instruction there was often neglected in favor of religious exercises. Consequently, my father called for a Protestant private school and found an excellent teacher, Mr. Vierkoetter, who provided instruction for the children, ages 5 through 14 years, in a single room. There was no strict division [of the children] into classes. Nevertheless, the instruction in all elementary subjects was excellent, with the result that my sisters and I were advanced in learning upon entry to other schools for older [children]. The teacher even went so far as to give instruction in Euclidean geometry to the more gifted pupils, both boys and girls. In later years, much amusement resulted when my sister Fanny, whose husband was a wood handler and unexpectedly needed to extract a cube root for his business, managed to solve the problem according to the knowledge given her by Mr. Vierkoetter.

The good instruction was made possible, to be sure, by the small number of pupils, which scarcely exceeded 20; for the school was originally attended only by children of the few Protestant families and several Jewish children. Only later did a few enlightened Catholics ask permission to be allowed to send [their] children there. This was a slightly hazardous undertaking; for the contrast between the two denominations was considerable, a fact which was frequently made clear to us children in quite unpleasant ways. We Fischers, as Protestants, were known as "blueheads" or "Calvinist calfheads", and we were obliged to engage in fighting when the excess number of Catholic boys made a successful defense seem unlikely.

On another occasion, when we were not split up but could appear as a unified force, success was our portion. My cousin Lorenz Fischer generally played a leading role in these heroics. He later performed outstanding deeds in the war of 1870 as well as when hunting as a private citizen.

When I was nine years old, the teacher Vierkoetter gave up his post in Euskirchen, because he had been offered a very lucrative position as Inspector at the reformatory in Brauweiler. Consequently, I transferred to the higher middle-class school in my hometown, which, like the *Progymnasien* at the time, offered four years of instruction in Latin, Greek, and French. This school was housed in rooms at the monastery church and fell under the leadership of Catholic priests. The rector, Chaplain Heine, was a prominent personality, tyrannical, irascible, but in spite of all this a genuinely good teacher. He worked to encourage his colleagues and thereby brought the school to an equal level with the state-run *Gymnasium*.

The handling of school discipline was capricious, to be sure. One injustice befell me, which made me aware for the first time of the great value of paternal protection.

A classmate known for brutality, by the name of Flecken, ripped a toy away from me without any reason and flung it into the mud. I answered this provocation by ripping off his cap and flinging it into the same mud. He defiantly left it there and then reported my supposed misdeed to the rector Heine. My excuse, that I was the injured party, was scarcely heard, and I was ordered to return the cap in its proper condition. This was not possible, however, as it had in the meantime been purloined by an unknown hand. Consequently, I next received detention, so that I had to remain at school the entire day and was not permitted the midday meal. The next day, when the cap was still not restored, the rector sent me home with the comment that I was expelled from school.

At this point my father felt it was time to take action. The next day he sent me back to school with a letter for the rector. I have never learned the contents of this letter, but it must have contained many a powerful word concerning arbitrary treatment of youthful disputes and of an appeal to the state government, should the rector maintain his entirely unlawful expulsion order.

The effect of this letter was actually astonishing. I was permitted to resume my place quietly in the school, and I enjoyed good and fair treatment from that point on. Yes, and I can even state that later I was on solid footing with the rector; for he loved mathematics, as did I, and he was happy to help me with difficult problems. I have retained a more pleasant memory of him than of most of the other teachers in my time at grammar school.

Not yet thirteen years old, I completed the *Tertia* [fourth form], the last form of the school, and was discharged with a good [school] report. Now I had to transfer to a *Gymnasium*. Previously, all sons of the family who wanted to pursue higher education entered the *Gymnasium* of Duisburg, which at one time had been a university city. This was the case because [the school] was under Protestant direction, enjoyed a good reputation, and was also not too distant from our home. Unluckily, though, things had not gone well for my cousin Ernst Fischer, who recently had been in the *Secunda* at this *Gymnasium*. He was caught in a small violation of school rules and was sentenced to a comparatively long term in the lockup. This punishment also seemed too severe to our parents; he was removed from the school, [but] our entire family fell into disrepute.

It would have been simplest now for us to choose a *Gymnasium* in Cologne, Bonn, or Aachen, but my parents, and particularly my mother, placed value on attending a Protestant school. And so, the choice fell on Wetzlar, even though this city was a full day's travel from my home. So as to be sure, and to indulge our strong craving for independence, our parents advised us, on their own accord, to have a look at Wetzlar and search for suitable quarters.

And so it was that during the autumn vacation of 1865 I traveled to Wetzlar, accompanied by my cousin Ernst Fischer, who was four years older, and by my cousin Lorenz. We traveled via Cologne to Wetzlar. The ancient city was pleasing to us. We also soon found quarters that we considered good and, most importantly, liberal. The landlord leased most of his house, both to students at the *Gymnasium* and to soldiers who had signed on voluntarily for a year to the local hunting battalion.

After the rapid conclusion of our business, we joyfully continued our journey via Giessen to Frankfurt. This old city made a great impression on us, with its many sights and historical remembrances. But our stay took an abrupt turn, brought about by a visit to a boarding school for girls, where my youngest sister Mathilde, as a seventeen-year-old adolescent, was being educated to become a young lady. She was beside herself with joy to see [her] brother and cousins again so unexpectedly. She assailed us immediately with a request to take her to the circus, which was in Frankfurt at the time. After some resistance, the head of the dormitory gave her permission, on the condition that a second young lady and a teacher, as chaperone, also be taken along.

We gallantly took on these stipulations. However, the expense was too great for our modest travel budget, and the next morning we hastened to leave the beautiful but expensive city. Still, our remaining money was enough for the cheapest seats to Mainz, and from there by ship down the Rhine back home via Bonn. It was my first extended travel, lasted seven days, and left me with so favorable an impression that the love of travel has persisted into my old age.

Soon thereafter came the departure from [my] parents' house. For many years to come, I would return only during vacations from school. The happiest part of [my] youth was over, for I have never had it so good as under the protection and in the cheerful atmosphere of [my] father's house.

Despite the prosperity that stemmed from successful operation of the business, our material life, though quite comfortable, was simple. We were spoiled only through the good, robust, and tasty nutrition, the preparation of which in the Rhineland was given greater weight at the time than in most other parts of Germany.

Also, the wine flowed freely in our house, often visited by guests. But it was a strict rule that we children under 14 years were not allowed to enjoy any alcoholic drinks. Above all, family life offered a cheerfulness that depended on the spirit of my parents and their happy marriage.

Chapter 2
Parents

My father Laurenz Fischer was born November 4, 1807 in Flamersheim; consequently, he was nearly 45 years older than I. His birth announcement is drawn up in French; for at the time the left bank of the Rhine lay under French administration, and the functions of the current registry offices had already been transferred to the "Maire."

Flamersheim is a small place in the foothills of the Eifel Mountains, yet still in the fertile plain, perhaps 7 km from Euskirchen. It had an old Protestant community with the see of the pastor, who, up to my time, also helped care for the small [Protestant] community in Euskirchen. The Protestants [there,] just as everywhere throughout the Catholic Rhineland, were more educated and more industrious than the Catholic masses and consequently also enjoyed a comparatively high standard of living. The business concerns of Flamersheim had also lent it significance well beyond the number of inhabitants. Furthermore, the largest property in town, the so-called *Burg*, with its stately main house that resembled a castle, its lovely park, and rich estates, remained continuously in the possession of a Protestant gentleman.

According to tradition, the *Burg* had been [part of] a dairy farm of Charlemagne, supposedly of many steadings, which had their special assignments and thereby lent their names to the surrounding villages, as Schweinheim (pigs), Stotzenheim (mares), Roitzheim (steeds), Buellesheim (bulls), Kuchenheim (cows). Whether this interpretation is historically justified or a product of the lively Rhenish imagination, I can't say.

Suffice to say, this Flamersheim was for several centuries the seat of my forebears. The oldest named in the church books is Johann Fischer, who died in Flamersheim in the year 1695, almost at the same time as his wife, who was born Hermany.

The paternal side was nearly extinguished with my grandfather, for he married only at 49 years. The story goes that this marriage was a lucky occurrence. He [had] lived together with an unmarried sister who openly tyrannized him. One day she denied him the key to the wine cellar, because he had unexpectedly brought home a guest. At this point he resolved to make himself independent through marriage. His

© The Author(s), under exclusive license to Springer Nature Switzerland AG 2022
D. M. Behrman and E. J. Behrman, *Emil Fischer's "From My Life"*,
Springer Biographies, https://doi.org/10.1007/978-3-031-05156-2_2

choice fell on Miss Helene Conrads from Mülheim-on-the-Rhine, likewise from a family of merchants.

My father was the second child. The two last, my uncle August and aunt Elisabeth came into the world in 1812 as twins. Their birth cost the mother her life. Also, the father died comparatively early at 61 years from a lung inflammation. The five orphaned children would probably have had a sad lot in life, if the unmarried siblings of [their] mother, two brothers and a sister, had not adopted them.

The oldest [sibling], Uncle Hermann Conrads, gave up his comfortable business in Mülheim and moved to Flamersheim so as to support the children of the paternal business. He kept the two youngest children (twins) with him. The three other brothers Friedrich Arnold, my father, and Otto Fischer came to Mülheim and were brought up chiefly by [their] aunt.

This woman played a large role in the memories of my father's youth. She was unusually gifted, socialized only with men, preferably disputing with academicians, particularly theologians. She had become well-acquainted with Latin and desired late in life to learn Hebrew, so that she might read the Bible in the original, although as an intellectual child of the French Revolution she thought purely atheistically. She directed the business in Mülheim and [was] no less the authority at home, where the good Uncle Heinrich, an unusually strong man, had nothing to say. The three boys who were entrusted to her protection were by no means spoiled but were soon inured and obliged to [develop] independence. When they were grown and married, the aunt persisted in addressing them as "boys" and would strongly dispute any disagreement with such styling or [with] her definite opinions. Nevertheless, the boys were devoted to her in gratitude, just as they adored both uncles as protectors.

My father and his older brother Friedrich Arnold were determined from the outset to be business[men], which was, particularly in my father's case, well-suited to [his] disposition. He did not have any great success in school; consequently, he left school at 14 years and was apprenticed in a commercial business to a Mr. Moll. Here he soon had the opportunity to act as a traveling salesman, which suited his inclinations [then] and throughout his 70-year business career. In his later years he often joked that he had learned more in the tavern than in school. Nevertheless, he was clever enough to recognize the deficiencies in [his] schooling; therefore, he later acquired considerable knowledge through private instruction in French and in mastery of the German language, as well as in legal and business matters. In his later years he made an attempt to learn English, for he had to make frequent trips to London to buy wool. But this [attempt] was too late; no one could understand his English.

At that time, Prussia already had an arrangement for [young men] to serve a one-year term of voluntary military service, but my father failed the exam. A putative physical injury, which certainly was no more than an incorrect medical diagnosis, obviated the requirement for three years as a soldier. In reality, he was a man of unusually robust constitution and health, to be discussed later.

After [my father] turned 18 years old, Uncle [Hermann] Conrads transferred the paternal business in Flamersheim to him and his brother Friedrich Arnold and moved back to Mülheim-on-the-Rhine. The two young people were well aware of their responsibility and seized the task, seemingly difficult for [such young men,] with great diligence and business acumen. It is a fine mark of sibling unity, that profits from the business would go not just to the two brothers but equally to all five siblings.

The third brother Otto was an excellent student, evinced an early inclination toward study, and, after completing his studies at the *Gymnasium*, became a medical student at Duisburg. After taking the state medical exam, which at the time had to be taken in Berlin, he went to Paris for one year and decided to become a surgeon. Not long after he became the leading medical doctor in the surgery department of the city hospital in Cologne. Later he turned down an appointment as Professor of Surgery at Bonn, because the resources and medical supplies seemed inadequate there. In his position at Cologne, where he practiced for over forty years, he earned his reputation through brilliant operations and exceedingly careful [post-operative] treatment of patients. For two decades he was the most sought-after doctor in the lower Rhine. His help was also in demand in Holland, Belgium, and France. An eccentric personality, he was a man of his region, in the best sense, about whom the active imagination of the poor people of Cologne wove into a succession of legends.

The fourth brother August also received a good education in school. After he completed his training in business, he served one year in the *Pionier* batallion in Cologne to fulfil his [military] service obligation, then likewise entered the business in Flamersheim.

About this time the twin sister Elisabeth married a Mr. Dilthey from Rheydt. He had originally intended to become a theologian, but the untimely death of his father necessitated that he take over, with his brother Wilhelm, the paternal business in silk and velvet fabrication. From this marriage stemmed the numerous Dilthey family in Rheydt and the surroundings. Of the six sons and three daughters, only the jurist Richard Dilthey and one sister remained unmarried. The others celebrated numerous descendants, and when my aunt died at the age of 87 years, there were perhaps forty grandchildren. The old woman was the center of her family until the end and enjoyed the admiration of all due to her prudence, her energy, and her kindness.

The business in Flamersheim prospered through the combined efforts of the brothers. My father traveled extensively on business in order to increase sales and to expand the business into new items. He often related with satisfaction, that only a few years after the discovery of the quick vinegar process, he established an operation in Flamersheim for that purpose and earned much money thereby. At that time, he also added a small malting house to the business. But he became convinced, as a result of these expansions and from experiences while traveling on business, that Flamersheim was not the right place for a large business, and he repeatedly pondered the plan to transfer it to Cologne. His brothers were entirely in opposition, [both] from aversion to the big city and the increased risk. He managed to get accepted a move of the main business to Euskirchen, which just at that time had acquired good roads to Cologne, Bonn and the Eifel. Thus arose, albeit gradually at first, the building complex I have previously described.

The two main houses, if I am not mistaken, were erected in 1835/1836. From that time on Euskirchen was the headquarters of the firm Fischer Brothers. Only Uncle Friedrich Arnold remained in Flamersheim, but the old company lasted another thirty years and expanded to other businesses, for example, the purchase of a beautiful forest near Flamersheim.

My father and his brothers complemented each other exceedingly well. He didn't like the detailed work in the counting house and warehouses, but this [work] was performed by the other two with great care and [business] acumen. On the other hand, he was glad to travel on business, not only to make sales, but also to collect debts and to search out new business opportunities. He jokingly named himself the Foreign Minister. Naturally, his duties included representing the firm in court, at government offices, at public sales, etc. As a rule, he took the initiative when new enterprises were begun, among which the previously-mentioned purchase of the forest stands out.

The community of Flamersheim had, in the neighboring foothills of the Eifel, a beautiful alpine forest, consisting mostly of oak and beech, from which [business] concerns at the time drew only a small income. Members of the community had a so-called claim, which gave them rights to timber and firewood, which at the time had but little value. Accordingly, the general feeling shifted to a division of the community forest. However, it was not possible to implement [this change] until after the uprising against the sovereign government in 1848/49. The forest was then publicly auctioned and the cash proceeds divided among the individual claimants. My father purchased a sizable piece for the firm, perhaps 750 ha.

A few years later the price of oak wood climbed unusually high, due to the building of the railroads. Consequently, [much of] the forest was soon cut down, the beautiful oak logs turned mostly into cross-ties, and the bark harvested for tannin. My father strove for proper management of the beautiful property by hiring a forest ranger.

We younger members of the family also enjoyed the forest. Wildlife was abundant, namely deer, sows, and woodcock. For convenience, a small hunting lodge was erected on the beautiful *Heidberg* with an impressive view of the Eifel mountains. From here we undertook many an outing, mostly for hunting, and were also able to spend the night. A well-stocked wine cellar provided cheerful drink. Our forest ranger was a superb cook, who knew how to turn the meat dishes brought mostly from Euskirchen, together with fried potatoes, bread-and-butter, and coffee into a delightful meal. Battues, which took place in November and December, were the high point of the hunting season. My father extended many invitations and hosted a merry meal afterwards at a guest house in Flamersheim.

After owning the forest for over fifty years, my father sold it again, at a handsome profit, to the Haniel family, because no one in our family could manage the property any longer.

Another enterprise, which was a contributing factor in my choice of career, was the acquisition of a small wool-spinning plant in Wisskirchen, a village about 4 km distant from Euskirchen. My father, who knew nothing about this industry, succeeded in finding in Aachen an excellent master of spinning, Mr. Allmacher. The firm spared no pains, equipping the factory with the best machines and replacing the [original] water power with steam. Consequently, the yarns produced were soon among the best in the district. Their use was not limited to the cloth industry in Euskirchen, but they also went to the Munich-Gladbach-Rheydt district, where they were used in the production of half-wool goods.

The purchase of raw materials, above all the wool, was naturally attended to by my father. Originally, [wool] was had from sheep-farmers in the area, particularly in the Eifel. In my early years I rode along here and there on many a trip [to buy] wool. In the poor villages of the Eifel business was first agreed upon with the pastor, after which the farmers would appear with their wares. The wool was appraised, weighed, paid for in cash, and loaded on the wagon. However, soon the local sources of wool proved inadequate for the enlarged factory. Also, the local wool was too coarse for finer cloth.

At that time wool began to be imported, chiefly from South Africa and Brazil. German merchants handled this trade, naturally with a corresponding markup. The market for this wool was London, where it was sold at large auctions. As soon as my father became aware of this, he traveled to London, where, with the help of an agent, he fulfilled his requirement for wool. He was among the first manufacturers in Euskirchen who obtained their raw materials directly. For a further reduction in the cost of yarn, a dye-works would be necessary. My father also had this in hand, but he had run into great difficulties.

Aniline dyes were little-known at the time, and processes of dyeing with natural dyes such as indigo, madder, yellowwood, and logwood were purely empirical. There were a few essential books for practical dyeing, and it was also possible to purchase formulas from dye[masters]. However, in practice all of this usually failed, and the management of an indigo vat was seen as a difficult art, that only through many long years of practice could be mastered.

It was therefore understandable in the newly established dye works in Wisskirchen, about which our otherwise so admirable spinmaster knew nothing, that many failures occurred, which led not only to losses but also to exasperating discussions with customers.

My father, who had always a high opinion of his own enterprise, therefore started a small experimental dye-works in Euskirchen, where he personally tested formulas purchased from various dye[masters], not omitting variations, particularly with the mordant. But in conducting these experiments, he soon realized how troublesome was his lack of chemical knowledge. He often remarked that if one of us boys would study chemistry, then all difficulties would easily be overcome. His respect for chemistry continued to advance as he came in contact with the blossoming industries along the Rhine, particularly the manufacture of iron and cement. I mention this in detail, because it influenced my choice of career, as it did for my cousin Otto.

The businesses previously discussed were all underwritten by and managed for the benefit of the firm Fischer Brothers. However, as my father's desire for ventures grew along with [his] wealth, so [my] more conservative uncles, as they aged, wished to avoid the ever-greater risks. Although he remained a member of the firm, [my father] busied himself with other ventures on his own.

Together with some businessmen from Cologne, he purchased at public auction larger estates whose owners had fallen into bankruptcy due to negligent husbandry, then sold them piecemeal to farmers in the surrounding areas. The times were quite favorable for farming in the Rhine region, and most buyers could pay for the newly-purchased land from their earnings in a 9-year time frame. [Payments] were generally made in cash, silver, which the farmers personally brought, not to the firm's counting house, but to our dwelling house. As my father was often away on business, this [duty] was left to my mother, who performed the bookkeeping with such promptness that would be a credit to any clerk.

My father had such extensive practice in the valuation of real estate, as well as in the purchase and sale [thereof], that one could place complete trust in his judgment. He quickly recognized the general economic danger posed to the Rhineland as overseas competition in grain farming expanded. He would often say, with sincere pity for the farmers: "There is nothing to be earned any more in farming." Already at that time he considered state support in the form of tariffs on [imported] grain necessary.

His interest in ventures therefore turned increasingly to industry. Some small mining ventures in the Eifel which focused on obtaining iron ore brought him little joy. The same was true regarding the financial prospects for a partnership in a cement factory in Obercassel near Bonn. This [factory] was established by Dr. Bleibtreu, an erstwhile pupil of A. W. Hofmann in London. He had learned in England about the manufacture of Portland cement and brought to life the first factory [for its production] in Germany, in Stettin. A second such factory was the plant in Obercassel. This [venture] would probably have been as profitable as the one in Stettin, if it were not linked to other unprofitable ventures. At that time the income of the public company *Bonner Bergwerks- und Hüttenverein* was quite modest, but my father seldom missed a meeting of the board of directors in Bonn, as he always met there with a succession of good friends from Cologne and Düsseldorf and followed the often-exasperating meetings with a merry meal at the *Stern* guest house. When I attended the *Gymnasium* at Bonn, I was sometimes allowed, to my great pleasure, to take part. The chemist, Dr. Bleibtreu, who was not only good at making cement, but was also a master in the preparation of peach punch, always advised me on these occasions to become a chemist.

The most prominent personalities on this board were my father's friend Albert Poensgen, a manufacturer in Düsseldorf, and a Mr. Muehlens from Cologne, a manufacturer of eau-de-Cologne [and] an outstanding joker.

One day when two of my sisters were present for the board-of-directors meal, [members of] the student fraternity *Borussia*, to which both sons of Bismarck belonged, were also there. These gentlemen seemed not to be very hungry, for they began their meal by drinking champagne and lighting cigars. This seemed too free and easy for the old Rhenish gentlemen, and in a resounding voice Mr. Muehlens gave the headwaiter the assignment to inform the young men that it is not the custom in these parts, when in the company of women, to begin a meal by smoking tobacco. The waiter carried out his assignment, whereupon the entire fraternity rose and marched out of the room, spreading clouds of tobacco [smoke]. Mr. Muehlens watched them with amusement and accompanied the [exit] march with the words "very well done." My father also was inclined to censure openly such violations of etiquette.

At the end of the '60 s my father invested a large sum in a business that caused him much worry and vexation, but later also great satisfaction. This was the establishment of a brewery in Dortmund. The initiative came from an engineer, Heinrich Herbertz, who owned a large coking plant in Dortmund. He noticed that the other breweries in Dortmund were prospering and concluded that [he] could make another good business here. He was related to my brother-in-law Mauritz and his brother Heinrich.

After my father had made the necessary inquiries, he agreed to the proposal, and so the brewery Herbertz & Co. was founded, in which my father originally was the principal partner. From the outset it was a good company, for the beer brewed there soon enjoyed a good reputation, and sales rose with the rising economy, particularly after the German–French war. Unfortunately, the enterprise was converted to a publicly traded company, and it fell on hard times due to bad management. Reorganization [of the company] was necessary, and with a complete change of leadership it once again became an orderly operation and soon returned to profitability.

The stock is still now, for the most part, in the possession of the families Fischer and Mauritz, and the enterprise has had quite good success the last 40 years.

My father was chairman of the board of directors for several decades. He also handled the purchase of hops for the brewery. During the time of my residence in Erlangen, I made many a trip with him to the hops market in Nürnberg. He knew how to assess the greatly-variable quality of the hops with confidence using empirical tests such as color, smell and feel while avoiding being taken in by the dealers. Also, he was otherwise greatly interested in the well-being of the brewery, particularly in the technical aspects. He always liked to say: "Above all else, the product must be good." At the laying of the cornerstone he even couched his wishes for the future in a pretty verse: "Brew good beer/I advise you [here]." Through his agency, I also had several opportunities to render small services to the brewery.

In the first year of my residence in Munich I fortuitously heard of new ice machines of Professor Linde and their use in the Sedlmeier brewery. I related this to my father when visiting on vacation. He immediately assigned me the task of gathering more precise information. This turned out very favorably; it was confirmed through the firm Meister, Lucius, and Brüning of Höchst-on-the-Main, who were using a Linde machine in the production of azo-dye materials. As a result, the Dortmund brewery immediately made contact with the company founded by Linde and, to the best of my knowledge, was the first north-German brewery in possession of a Linde machine and later also succeeded in installing a very efficient basement cooling [system.]

Perhaps no less important was a lesson on the fermentation process I gave to the brewmaster. During the winter semester 1876/77 I was staying in Strassburg and, through Dr. Albert Fitz, became aware of Pasteur's book "Études sur la bière" that had recently appeared. The brilliant researcher had recounted his experiences concerning contamination of the brewer's yeast by other microorganisms and the damaging effect on the quality of the beer. When I reported this to my father, he urged me to study the material thoroughly, and, as this also was of scientific interest to me, I agreed to his request. An excellent microscope was immediately procured, and, with the help of Dr. Fitz and the botanist Prof. de Bary, I started a study of molds, yeasts, and bacteria, which later would prove useful to me in the work on sugars. Next, I had to make practical use of the new knowledge. Consequently, I moved to Dortmund with my microscope for several weeks in order to make the new [scientific] achievements known to the management at the brewery.

Probably I was the first chemist in Germany to make such an attempt, and I must confess, I ran into great mistrust among these men of experience. [They] made every attempt to lead me astray, particularly with false assertions concerning the origin and condition of the types of yeast to be proved. However, when I was easily able to discover the contaminated types, with the help of the microscope, they grew more serious. I did not succeed in training [even] one of the men in correct use of the microscope, which [might have] established enduring control of the yeast. However, my instruction on how good yeast and the consequent beer can be spoiled made an impression. For example, adjacent to the cooling room, where the finished wort was cooled in open air, was the horse stall and a substantial manure pile. A change was soon made at my suggestion. Also, [my suggestion that] cleanliness of all vessels that touch the cooled wort and the beer be maintained, which I preached as particularly important, was to the liking of the reasonable brewmaster, because this was in accordance with his practical experience.

Perhaps 30 years later, at a jubilee of the Institute for Fermentation Business in Berlin, I was named an honorary member. The leader of the institute, Professor Max Delbrück, took the opportunity to emphasize that the choice was motivated not only by my chemical work, but also, jokingly, by my efforts at the Dortmund brewery, which demonstrated an interest in the practical [aspects] of the brewing business. The work of Pasteur that I tried to make known [at the brewery] had, in the meantime, been extraordinarily improved through studies by Professor Ch. Hansen in Copenhagen, whose work became fundamental in the brewing industry. He was also at the jubilee and was also named an honorary member.

As previously mentioned, my father was a member of the board of directors of the publicly traded brewing company and for many years held the office of chairman. Other publicly traded companies also named him to their boards, for example, the *Bonner Bergwerks- und Hüttenverein*, the *Röhren- und Eisenwalzwerk Poensgen* in Düsseldorf, a glassworks in Stolberg, a boiler shop in Kalk-on-the-Rhine, and the Concordia insurance company in Cologne. When, due to increasing deafness, he announced his intention to leave, the always cheerful, funny old man was asked to stay on. And so it came to pass that after the ninetieth year of his life, he could boast of being the oldest member of a board of directors in Prussia.

As the preceding portrayal of his business ventures has shown, my father was a versatile, clever businessman who correctly assessed the opportunities of [his] business life and seldom entered a poor venture. The deficiencies in his education were later compensated by his practical experience. He was not a quick thinker, but after he had thoroughly studied an opportunity that lay within his circle of experience one could be certain that he had fully grasped it and had foreseen the consequences.

I have known from experience that he dictated contracts or conditions of sale for public auctions to legal secretaries, because their own drafts were not clear enough. The last example of this type [that] presented itself was his testament manuscript, which showed the same clarity of form and of thought as all his pieces of writing. Shortly before his death, according to his wish, it was converted to a legal document by a legal secretary in Berlin. When I then read this document aloud to my brothers-in-law for the division of the estate, they declared as one: "Granddaddy didn't write that; he never expressed himself so unclearly." The confusion had arisen due to the legal secretary's adaptation and legal turn of phrase. Luckily, I was able to set aside all doubt about the sense of the legal document by producing the manuscript.

Along with his clarity of mind was an unusual physical vigor, which had not escaped the sharp eye of his mother. As he liked to relate, she had reported in a letter to her sister Conrads about the little Lor: "He seems to be a bit stupid, but it is a genuine pleasure to watch his powerful little body." It is therefore also no wonder that he was fond of all physical activities. Riding, dancing, gymnastics, and shooting were all well-known to him. He hunted into his old age. [His] hand and eye had remained so capable that at 93 he was able to send me a hare he had shot himself. Along with this came a great cheerfulness of mood, which [even] severe losses would disturb only temporarily. His needs [included] daily walks in fresh air and 1 to 2 h of company in the evening at the guest house or casino with a glass of wine and a cigar or pipe of tobacco. And when he returned home his gaiety was infectious for the entire family circle. The laughter in our house was often so loud and sustained that passersby on the street stopped in astonishment.

Of course, he practiced hospitality with gusto, and my mother sometimes had trouble providing for the large number of guests who suddenly gathered. Family celebrations were particularly lofty and merry, for example, at my sisters' weddings. No fewer than 7 [weddings] were celebrated in our house, six for my sisters, of whom one married twice. And when the time came for the little daughter of my uncle in the neighboring house, the celebration, out of habit, was also held in our house. The uncle just had to bear the moderate expense. On such occasions my father let flow all wellsprings of his gaiety. Although he was not a gifted speaker, nevertheless his after-dinner speeches, which sometimes were adorned with small, amusing verses, always aroused great jubilation.

His Rhenish sense of humor and his love of jokes also led him fairly regularly to the Carnival celebrations in Cologne. In his old age he once invited a large crowd of relatives and friends to a guest house in Cologne. No one knew for what reason, until the host solved the riddle in his after-dinner speech. "All the world is celebrating now," he began, "and so I thought I must organize a celebration, for today is the fiftieth time that I have taken part in the Carnival in Cologne." One can well imagine the high spirits that prevailed at this celebration.

Even more characteristic and, for those not from the Rhineland, difficult to understand, was a Carnival idea he carried out shortly after the death of his brother Otto, the medical doctor in Cologne, whom he deeply respected. Although he was nearly 80 years old, he did not believe he ought to abstain from Carnival. Therefore, he attended the masked ball, but with the mark of mourning, in the costume of a Moor. This was limited, though, to blackening his face with a burned cork.

Unavoidable things, such as the death of his four siblings, with whom he had lived in true and ever-benevolent friendship, and who all preceded him in death, he quickly overcame. Much more difficult for him was the death of my mother, with whom he had been married 46 years and whom he outlived by 20 years. After her death he spent another 10 years in Euskirchen, then suddenly he was prompted, under peculiar circumstances, after 57-year residence, to give up his abode.

In the year 1892, the obligatory self-disclosure of income was introduced for the Prussian income tax. In the Euskirchen administrative district it turned out that my father had the largest income in the district, which was unexpected, given his simple lifestyle.

The increased income tax was, for the usual Prussian assessment, also for the benefit of the local government; and the percentage at which local taxes had previously been assessed was supposed to be sensibly reduced. This had also been suggested by my father; but instead, only the business tax was reduced. When my father referred to this as unjust and hinted at the possibility that wealthy people who no longer operate a business could leave the city [to avoid] the rather high local taxes, the retort came: a man such as he, at the age of 84 years, would no longer be able to make up his mind to change his residence. This reference to his age-related weakness angered him, and, in order to prove the opposite to people, he immediately decided to leave Euskirchen. Without saying a word about it to [even] one of his children, he dissolved his household, left the city and Prussia, and moved to Strassburg-in-Alsace in the summer of 1892. Shortly before [the move] he paid me another visit in Würzburg, to be discussed later. At that time, it was already too late to change his decision.

The resulting move took place in simplest form. At the train station in Cologne, a relative met the old gentleman in a most basic suit, shotgun over his shoulder, with the dog on a string and a modest small suitcase in the other hand. To the question: "Well, Mr. Fischer, where are you going?" came the terse reply: "Domicile change to Strassburg." Here he lived, first in a guest house and later with my cousin Ernst Fischer, the associate professor of surgery, who had established a private clinic in his roomy old house in the Cooper Lane. His wife, who came from an Alsatian farming family, maintained a simple but good household.

My father soon found a congenial circle of acquaintances, especially among hunters. After [only] a few months, he had participated in a hunt and, as he laughingly related, also already had a lawsuit, that he properly won. In later times he had contact with officers of the garrison and was invited to their battues. Due to his advanced age, he received the place of honor next to the highest-ranking general, and it gave him great pleasure when shooting hares to beat the general to the trigger, which by younger officers was not allowed. By the way, he was accustomed to intercourse with officers, as autumn military exercises very often took place in the surroundings of Euskirchen. Indeed, even the imperial field exercises were twice held there, and high-ranking officers were always quartered in our house. He also took part in deer- and sow-hunts in the Vosges.

He made frequent business trips from Strassburg to the lower Rhine [region] and also to Berlin, particularly for meetings of boards-of-directors.

After he turned 90 years old, he made up his mind to return to Prussia. However, regarding the taxes, he had, in the meantime, laid out firm principles that would not easily be attained. He would gladly pay the state tax, one often heard him say, because protection of the state is indispensable to everyone. However, he had little use for local governments, and he didn't see why he should pay them much. Therefore, he wished to choose a locality distinguished by low taxes. His first choice, in the summer of 1898, was Wannsee, near Berlin, where at the time only a 40% surcharge on the state tax was assessed. When the first tax bill came, however, he saw that an additional 30% tax was assessed, and immediately notified [the authorities] that he would leave.

The next choice was a small place in the Harz, where his eldest grandson Heinrich Mauritz was employed as the royal mining official. The arrival of the wealthy citizen seemed to the other members of the district as a favorable opportunity to satisfy all possible desires of the district, and the taxes went astonishingly high the following year. The old gentleman immediately disappeared and now found a home in Griethausen, a village near Cleve, where [people] abstained from exploitation of the migratory bird. He rented a room with the beadle and maintained this domicile until his death, although he scarcely spent 24 h there.

Instead, he lived by turns with his children or sons-in-law or grandchildren in Rheydt, Uerdingen, Berlin, Herdt or Dortmund, respectively, but staying nowhere longer than 89 days; otherwise, he would, according to law, be responsible for paying the local tax. The remaining part of the year, namely the summer months, he spent traveling and thereby became acquainted with as many [people] as possible. One day he wrote me, that at a guest house in Heidelberg he had accidentally met Robert Bunsen and had introduced himself as the father of a chemist. In fact, [Bunsen] was younger than he, but harder of hearing. Nevertheless, a long and interesting conversation between the two developed.

In the winter of 1900/01, he took part in the hunt for the last time, because his eyesight, which had been so good, was failing. This was due to a slight cloudiness of the lens, while the optic nerve had remained intact. Aside from that he still enjoyed good health into the spring of 1902, apart from the deafness. At my house he still dressed formally for company in the evening. But when the 89 days were up, he moved with the usual punctuality to Rheydt [to stay with] his son-in-law Arthur Dilthey. Here, it is said, he frequently idled away the hours in guest houses, mostly in the company of young people, paying homage beyond healthy measure to the pouring of beer and wine. His old heart was not up to it, and the first sign of [heart] failure was made known by swelling of the legs. The opinion of the doctor was that he should change his lifestyle and avoid the pleasure of alcoholic drinks, to which he responded: "How many people of my age have you treated?" He therefore persisted in his habits. Over the course of the summer the dropsy became worse, but this did not bother him greatly. At the beginning of October, as a very sick man, he came to me in Berlin accompanied by his grandson Alfred Dilthey. Without suffering greatly, on October 16, 1902, 18 days before his 95th birthday, he passed away.

One day before his death, he again left his bed for five to six hours and worked to complete his ledger, which he always kept with him when traveling. A later check of the book showed only a single error was committed on this last day.

On the day of his death, in a conversation with me, he reviewed his life and expressed himself as very satisfied. One hour before his death he drank yet another large glass of beer, apparently with satisfaction; for his last words were: "It is good fortune, that the common man can cheaply enjoy so good a drink." He died, as he had lived, as a perfect atheist, but in true devotion to his wife and children, to extended family, and to many friends.

Politically, he was earlier a Rhenish progressive, later a national liberal, and always in strong opposition to the ultramontanist party. Although he grew up in the capitalist age, still he had sense and understanding for the social movement of modern times. He wanted nothing to do with the future socialist state, but as for the ambitions of the worker he was fond of saying: "The common people are fully justified when they try to improve their lot." Publicly he did not stand out either as a speaker or as a writer. On the other hand, for many years he took part in local government as city councilor and as a member of many commissions, and when the issue at hand was for general economic purposes, as for example to advance the building of railroads in the region, he was generally a member of the delegation to be sent to Cologne or Berlin to work with the royal government.

In spite of his vigor in economic matters, in character he was good-natured and honored the principle of "live and let live". Towards his wife and children he was not only very kind, but also considerate and left the strict side of child-rearing to the mother. My sisters could get almost anything from him with just a word of flattery. Only in serious matters, for example, in choice of a husband, did he urge caution and reason, and he would surely have opposed, with great determination, wedding plans based on fantasy. He was lucky to be very satisfied with all his sons-in-law.

For myself, I do not remember ever having been punished by him. Yes, I scarcely heard an angry word from him. Yet he did sometimes express his disappointment that I showed no interest in affairs of business or in inheriting material wealth. But he peacefully let me go my own way and only spoke to others with regret that [his] son did not have the knack for accounting. Unfortunately, he did not live to see this impractical scholar receive the Nobel Prize in chemistry and later, through a few small inventions, earn a yearly income beyond what he had ever had.

My mother was fundamentally different from my father in character, outlook and attitude, despite their very happy marriage. She stemmed from the second marriage of the iron manufacturer Johann Abraham Poensgen in Schleiden (Eifel) with Wilhelmine Fomm and was born February 19, 1819. Borrowing from a very well-written history of the Poensgen family that has appeared in print, I [can report] that my maternal forbears were producers of iron and iron products for hundreds of years, so perhaps I inherited a knack for chemical and technical processes from her. It is a peculiar accident that now in my old age I have again come in close contact with the coal and iron industry through the Kaiser Wilhelm Association for the Advancement of Science and, in earlier times, have even been a member of the board of trustees of the wonderfully planned Kaiser Wilhelm Institute for Iron Research.

After the death of her first husband, my grandmother married the physician Dr. Fuss of Gemund in the Schleiden valley. Consequently, my mother spent most of her youth there. For the completion of her upbringing, she came to the educational establishment in the Herrnhuter community in Neuwied on the Rhine. It was there, most likely, that she derived the deep religious convictions that she held the [rest] of her life.

She was very clever and eager for knowledge, and due to extensive reading, she became rather myopic in her youth, which I alone among her children have inherited. She would no doubt today have become a learned woman, but in her time study among women was not customary. As she married at 18 years and brought 8 children into the world over the following 15 years, her time was occupied with other duties. She became a capable housewife and also assisted my father with business matters. She knew how to command respect, and no one among the children or the servants would have dared to disobey her directions. She was more serious than my father, but she could laugh heartily at his jokes. When she encountered crude or mean remarks, she gave such voice to her indignation, that everyone in her presence was obliged to behave himself. Everyone considered her a loving mother, full of solicitude for her children and later for their families.

[Despite] being deeply religious, she did not allow my father's lack of faith and his habitual derision of church and priests to drive her crazy.

She was just as independent in her political outlook. As an ardent Protestant, she honored Prussia as proponent of the evangelical faith in Germany. When the unpleasant conflict between the Prussian government [led by] President Otto von Bismarck and the House of Delegates raged in the years 1863/4, and nearly all Rhinelanders, including my father, belonged to the opposition party, she was entirely on Bismarck's side: "Now Prussia again has a minister who is worthy of her; and you are too stupid to understand this man." She spoke this way to her children, and she gave such lively opposition to my father in her political opinions that he jokingly referred to her as "Mrs. Bismarck."

In addition, she was quite a pretty woman, particularly distinguished by the luxuriant deep-black hair and the large, intelligent eyes. As the only son I enjoyed her special love and affection. She always encouraged my inclination toward scientific studies, but the career choice of chemistry disappointed her. She would have much preferred if I had become a lawyer or medical doctor.

In general, she enjoyed good health. When she was 59 years [old] she traveled to Meran to keep my sister Mathilde company during the winter. [Mathilde was ill and staying in Meran to improve her health.] From Munich I visited her during the winter holidays. I made the trip accompanied by my friend Dr. Tappeiner, whose father was the best-known medical doctor in Meran. As we drove in at night from Bozen in a special hired carriage, the chill was so intense that we were quite uncomfortable, and on arrival in Meran the temperature had fallen to 12°. At that time, I changed my optimistic opinion of the warmer climate in the southern Alps locations, and I confirmed this judgment by many a similar trip to the South. In Meran, the mountain behind the city was so encrusted with ice that one could hardly climb the peak with ordinary shoes. The next morning, in the guest house "Archduke Johann," I found the water frozen in the washbasin. The ill people, by the way, were better cared-for; they had continuous heating.

In the middle of February, my mother returned to Germany from Meran in order to celebrate her 60th birthday at home. She visited me again in Munich and was glad to meet several of my friends as her guests in the "Bavarian Court" hotel. But this long railroad journey with inadequate heating and, of course, no sleeping car gave her a serious case of bronchitis, which left her circulation hindered. This together with the emotional distress over my sister's fate probably triggered heart disease, which began in the spring of 1879 and, after 3½ years of truly severe suffering, led to her death on September 14, 1882. She died in Uerdingen at the house of my brother-in-law Mauritz under the painstaking care of my sister Bertha. She is buried, as is my father, in the small Protestant cemetery of Uerdingen in the Mauritz-Fischer family grave. The cemetery was earlier quite pretty, lying in an open field, but now, unfortunately, due to rapid industrial development, is surrounded by tall and unattractive buildings.

My sisters were previously discussed several times. The slight aversion that [their] brother earlier sometimes felt in resisting their efforts at rearing [him] was, over the course of time, turned [completely] around, and a truly friendly relationship developed between myself and my sisters and their husbands.

I scarcely knew the eldest sister Laura, as in the year 1858 she married a young businessman Friedrich Mauritz from Uerdingen-on-the-Rhine. Consequently, at age seven years I was granted the honor of being an uncle, which resulted in many a teasing from my age-group. During this marriage, a bond developed between the Fischer and Mauritz families that strengthened and deepened with time. My brother-in-law was a splendid, cheerful man, hard-working at his business, a coal supplier, and my father [developed] a very cordial relationship with him. In his old age [my father] liked to spend a month at his son-in-law's house. He started several businesses with him, for example, the brewery in Dortmund. He liked to travel with him and often allowed that Fritz was as dear to him as his own son.

As a boy I have myself spent time with the brother-in-law during school vacations. In the surroundings of the mighty stream, which was new to me, the peculiar business, and the extended family with many children, I was greatly entertained and experienced much joy. Unfortunately, my sister Laura, a healthy and strong woman, died soon after the birth of the third child Alfred, who now has a very highly regarded position as director of the brewery and city councilor in Dortmund and apparently is busy with many other things as well. She seems to have been the victim of an infection, the exact nature of which I was not able to determine. Five years later, in the middle of the 1870 war, the widower married my third sister Bertha, who gave her husband three more sons. Unfortunately, she also died early, of a lung inflammation, in 1888. My brother-in-law Fritz followed her in death perhaps 10 years later, which for my father in his old age was a painful loss. Of his 6 children only 2 remain, the brewer director, already mentioned, and the son Otto from the second marriage, who is an engineer at a large machinery factory in Nürnberg.

My second sister Emma likewise was married fairly early, to a medical doctor Albert Winnertz from Krefeld, whom she met through my brother-in-law Mauritz. The doctor was a very clever and obliging fellow. Unfortunately, [he] was ill and died of tuberculosis after 3½ years of marriage. The young widow returned to the parents' house in Euskirchen with 2 children, Hedwig and Clara, and remained there nearly 10 years. That is the reason why, of all my sisters, I got to know [Emma] the best and became especially close with her. She was pretty, engaging, and quite elegant. My father consequently always named her grand dame. She played piano rather well. I remember many an evening listening to her play for hours, because she performed mostly classical pieces on a good piano. I believe her play was the primary motivation for me to carry on with music.

In the year 1872 she decided, after a long hesitation, to enter a second marriage, with my cousin Carl Fischer, who had patiently courted her. So, she came to Rheydt, where two of her sisters already were married. I often visited her there, and she repeatedly came to Würzburg and Berlin. In later years we also traveled together extensively. She pursued friendship with [her] brother to such an extent that her husband became jealous and complained of the slight. She died in the year 1901 of typhus in Nassau, and my brother-in-law Carl outlived her 14 years. She suffered the pain of losing three sons in [their] youth. The four daughters, two Winnertz and two Fischer, have all married and remain dear nieces to me.

My third sister Berta was a truly odd person, at heart good-natured, but very moody as a girl. She was often disagreeable towards men, yet very capable. After she married my brother-in-law Mauritz at age 29 years, she became a well-known woman in Uerdingen because of her unusual skill in cooking and in training servants, as well as for her quirks. She was a true comrade, mostly for young people, for example, the sons from the first marriage and their friends. She played cards, drank wine, and took outings with them, just like a student. Her unfortunate early death has already been mentioned.

The fourth sister Fanny was powerful of body but [had] little beauty of the face. She distinguished herself in her youth with a penchant for gymnastics, dancing, and pistol shooting. In 1865 she moved to Rheydt as wife of the woodhandler Max Friedrichs and died there in the year 1912 at the age of 70 years. She was clever and just as capable a pupil as she later was a housewife. As already mentioned, she was able, as an adult, to extract a cube root using the mathematical knowledge given her by Mr. Vierkoetter. She also was known in the family as a poet. She could compose lovely poems for a [special] occasion or a put together a small festival, which many requested of her, in an unbelievably short time. Of her children, two sons, Ernst and Max, and two daughters, Helene and Adele, are still living. The sons continue to run the significant wood-handling business inherited from their father while the daughters each married out of town. More details on the brother-on-law Max Friedrichs will be given later.

The youngest sister Mathilde was only 4 years older than I and therefore was the closest to me in early childhood. She was a dear, very cheerful, quite pretty girl, intellectually not particularly gifted, but obliging and consequently welcome everywhere. Her bearing brought her much attention from the officers who were quartered in our house during maneuvers or during the 70/71 war, but she preferred to remain in the civilian [sphere] and married my cousin Arthur Dilthey from Rheydt. After the birth of her third child, she took ill and died in the autumn of 1879, in Rheydt, after a futile [attempt at] cure in Meran. [Her death] was a heavy blow, not only for my brother-on-law, who was left with two sons and a daughter, but also for my parents, and particularly my mother, whose own health was very gravely affected.

I have retained a grateful and warm memory concerning dear "Tilla." 5 years later my brother-in-law entered into a second marriage, with Frieda Weuste, and she took it upon herself to be a genuine second mother to the three children. The Arthur

Dilthey couple still lives in good health in Bonn. Of the three children, the youngest, Alfred, unfortunately fell victim of the unholy war in Russia in 1915. I developed a lasting friendship with my brother-in-law Arthur, of whom more will be related later.

In early youth I was nearly as close to the cousins in the neighboring house as I was to my sisters. The eldest, Heinrich Fischer, 5 years older than I, the only member of the family to remain in Euskirchen, was inclined as a boy to isolate himself from the rest of the circle. He remained unmarried and has taken on the characteristics of an eccentric more and more. He lives in our old house and enjoys running the old spinning business, though on a smaller scale.

Concerning the cousins Ernst and the youngest Otto Fischer, I will have much more to report later.

Cousin Lorenz has already been mentioned because of his outstanding character-istics as hunter and warrior. In school he was less proficient, and, due to a foolish lifestyle, he died of tuberculosis at the age of 35 years. [Like Lorenz] his younger brother Hermann was better suited to physical activities than to the intellectual. He was a very handsome man, a good gymnast and rider, and he moved to Cologne soon after the death of his father. He became the progenitor of a warrior family; for both of his sons Kurt and Walter became career officers. His daughters were distinguished by their beauty, and both are happily married and living in greater Berlin.

The three sons of my uncle in Flamersheim also came in contact with us often through frequent visits, [made possible] by the small distance from Euskirchen. The two oldest, Karl and August, became businessmen and later led a cotton yarn business in Rheydt. There they also died. As already mentioned, Karl became the second husband of my sister Emma. Despite his somewhat sketchy education, he became a clever and skillful businessman. [Many] sought his advice in practical matters, and I frequently spent a part of my Easter vacation with him in Territet on Lake Geneva.

The second son August was of unusual physical strength, an outstanding hunter, and good-natured in company with an inclination to laughter. In business matters he deferred to his brother Karl.

Very different from his brothers, the youngest Flamersheim cousin Julius [had] a very bright and critical mind. He had inherited both the mental agility and peculiarity of his father. He became a lawyer, chose the career track to become a judge, and died comparatively early from untreated diabetes in Cologne, where he was a judge on the Higher Regional Court. He was married at Cleve, an earlier domicile. From this marriage came one son, who married the daughter of a large[-scale] sheep farmer in Australia and found a new home there. I hope that through naturalization and the protection of his father-in-law he has escaped the [prejudice] that our countrymen face in the English colonies because of the war. The only daughter married a Hans von Eicken from Hamburg, the son of my cousin Helene née Fischer from Cologne.

Cousin Julius not only had a gift for intellectual work, but he also had a talent for making inquiries. Even as a boy he knew everything that happened in the local area and surroundings. He was also interested in business matters. One could hardly hear something more peculiar than a conversation between him and his father. [The latter] thought very highly of his son's gifts and often gave his opinion in confidential circles that "the Jul will become a minister." In his old age the father's acquisitiveness and parsimony grew to such grotesque levels that he would have suffered material want, had not the son Julius [and] his brothers secretly borne the main costs of the household. Here is a typical story.

One day in his little office the father was missing half of a hundred-thaler note, and he suspected that a small dog, the favorite of his son Julius, had eaten it. He immediately decided to have the animal cut open and had already called for a surgeon, when Julius received word of the danger to his [dog.] He succeeded in mollifying [his] father by buying the dog for 100 thalers. Luckily, the transaction thereafter could soon be canceled when the second half of the unlucky hundred-thaler note was again found.

I got closer to cousin Julius during my time in Bonn as a student. At that time, he was a probational lawyer for the district court. Now and then we would share a bottle of wine at the *Rheinischen Hof*. I always had to wonder at his extraordinary talent to perceive an issue, which was completely clear to him, from a different point of view and to develop possibilities that a layman could not possibly conceive of. This skill, to see the world in a particular mirror, was very attractive to me. In the autumn of 1874, after the doctoral exam, I invited him to travel [with me] to Switzerland. Though little inclined to travel, he agreed.

I have seldom laughed so much as on this tour. The sights of our departure, [such as] the scenic beauty of the upper Rhine valley and the lovely mountain peaks of the Oden Forest and the Black Forest, left him relatively unmoved. On the other hand, he precisely scrutinized the train's timetable and made a note of each minute the train was late. After boarding the train, he took off his tie, collar, and jacket and substituted a raincoat. This was, [according to him,] the best way to guard against dust and dirt, which he held to be unhealthy and unattractive.

Our first stop was Basel. He was interested in the city and the inhabitants, but not in the countryside or scenery. He proclaimed the following principle for travel: to view the public buildings from the outside, the mountains from below, and the guest houses from within. I have heard similar words from other traveling companions, for example, my friend Wilhelm Königs, but never was this principle so scientifically carried out as by cousin Julius. Towards evening he always had the desire to drink a glass of wine. In strolling through the city, he was focused on finding a certain wine bar. He didn't trust the travel book at all, likewise the advice of the hotel servants. Suddenly he found a man on the street with an unusually large and fire-red nose. "This [man] will know," he immediately said, and the information received from this gentleman in fact led us to the correct address.

From Basel we made our way to Lucerne, where we stayed in a guest house of the second rank, a true Swiss house. The cousin was particularly interested in making contact with the local population [and] getting to know their habits, institutions, and opinions. He soon succeeded in engaging regulars in conversation. All this was accompanied by the consumption of considerable quantities of wine.

From then on, the travel path was precisely determined by him, and every detail was also retained [by him.] [The path took us] by way of Flüelen, Andermatt and the Wallis to Lake Geneva. Next morning, after we had boarded the ship to cross Lake Lucerne to Flüelen, he fell into a deep sleep and did not awaken until near the end of the crossing. This despite superb weather, fine seats on deck, and countryside worthy of seeing. At this point he looked well-satisfied and said dryly: "The people must think I have been here often." At Flüelen we had the choice of traveling to Andermatt either with the post [coach] or by walking. Cousin Julius quickly chose the latter. It was a humid day, and the road was quite hot and dusty from the midday sun. We had sent our luggage with the post [coach,] which [would] leave later, and [so we] set off. The cousin was of the opinion that most comfortable way to carry the overcoat is to wear it. This he did, and it wasn't long before the well-nourished cousin, unused to strenuous exercise, was sweating. Unfortunately, just at that time a squadron of Swiss cavalry were performing exercises in the area. Small units passed by frequently and raised monstrous clouds of dust, which the cousin held to be very harmful, so he sought to escape [the dust] by leaving the embankment for the open fields. After 2 h under these conditions, he was so exhausted that I became worried and suggested that he stop at the next guest house, then take the post [coach] the rest of the way. He agreed and did not leave the post [coach] until we arrived at Lake Geneva.

I continued to travel the more beautiful parts of the journey on foot, particularly the stretch from Andermatt over the Furka [Pass] to the Rhône Glacier. Along the way I found myself in the company of a young American, whose practical approach to traveling particularly appealed to me. He had come over from California and had no luggage aside from a small bag for money, toothbrush, and soap. As soon as he arrived in a city, he bought new underwear and disposed of the dirty. He had already traveled around Europe this way for 2 months, seen much, enjoyed himself, and hoped to extend the trip for several more weeks. In the evenings I always met up with cousin Julius. He then recounted many amusing anecdotes of adventurous travel in the mountains by the post [coach] and of curious traveling companions. He even discovered a new physical phenomenon at the Devil's Bridge: an inverted rainbow. At first I didn't believe it, but later I convinced myself that at a waterfalls, where water is very finely divided into tiny drops, and with correct position of sun and observer, this event can in fact occur.

The scenic beauty of Switzerland was little discussed [between us.] Instead, [Julius] found Geneva interesting as the city of Calvin and as headquarters of the Swiss financial business. Here I also experienced an amusing scene with him. He maintained that the water in Switzerland is especially beneficial, but one must take it directly from a natural spring. This despite the fact that he never drank water at home. He would love, therefore, to drink water directly from such a spring. He did just this at a spring in Geneva. However, in order to reach the water, he had to lie on the walled parapet and swing back and forth to catch the stream in his mouth. The sight of the short, very well-nourished man lying on the parapet of a spring and drinking was so amusing that a large circle of observers gathered. When he was finished and caught sight of the assembled crowd, he said softly that this again shows that one can, with little effort, provide people with much amusement.

From Geneva we took the omnibus to Chamonix, and I made several small mountain tours. Cousin Julius stayed true to his travel principle and remained at the hotel; however, through assiduous inquiries of tourists, leaders, and other people, he became far better informed about mountain tours than the majority of the travelers. Naturally, he also inspected all guests of the hotel and reported one evening that there was an agitated German professor outside by the weather house criticizing the instruments and lecturing the public on the imminent change of the weather. I later learned that this was Rudolf Fittig from Tübingen, who was already familiar to me as an outstanding chemist from the literature. At that time, I would not have dared to introduce myself to him; however, several years later I entered into a close relationship with him.

The walking tours I took at Chamonix [included] the well-loved (at that time) crossing of the *mer-de-glace* and climbing of the *mauvais pas*. For hikers used to climbing mountains, it is a mere stroll, but people without experience and without the proper footwear could very easily slip and have an accident. As I crossed the ice, I had a certain fright as I recalled that my father 5 years earlier had undertaken the same path with 5 women, namely my mother, two sisters, and the two cousins Marie and Helene Fischer from Cologne. He related afterwards that those were the most terrifying hours of his life. The women seemed to be less aware of this danger; indeed, they had a number of remarkable and amusing experiences on their tour of Switzerland which were recounted at our house for a long time.

Cousin Julius's impression of Switzerland was more sober. After we returned to the Rhine valley, he maintained that it is not worth the effort to travel to Switzerland, for here there are mountains and water, and snow and ice in abundance in the winter. To my knowledge he never again crossed the border of the German empire. On the other hand, in my later years I have often and with pleasure returned to Switzerland, and particularly to Lake Geneva; however, I have never again seen the Chamonix valley.

Chapter 3
Gymnasium

In October 1865, accompanied by my father, my cousin Ernst and I drove to Wetzlar, where my father registered the two of us with the director at the *Gymnasium*. Ernst was accepted into the *Prima* without any trouble due to his exit certificate from Duisburg. On the other hand, I had to work through a test, because the school in Euskirchen was not seen as sufficient. I passed the test without difficulty and was started in the *Untersecunda*, which was consolidated with the *Obersecunda* in one class. Even today the laughter rings in my ears, that classmates raised when for the first time I had to speak in class and they heard my distinctive lower-Rhine dialect. Naturally the Wetzlar dialect, which sounds much like Frankfurt German, seemed just as unnatural and comical to me. However, I was soon able to manage in this circle, and even, after ½ year, [reached] the top of the *Untersecunda*.

The quirkiest personality among the teachers was the mathematician Elstermann. He would enthusiastically clean the well-chalked board with his long black robe and then run the dirty finger through his long, bushy hair. He was a good person and an excellent teacher, who had the difficult task of filling youthful heads with mathematical thoughts, [which he did] almost with sorrow. When he found, on quizzing, a single pupil with a lack of understanding or interest, he would deliberately give a different pupil within reach a small box on the ear. As his exasperation grew, the blows became stronger, and when the recipient looked angrily at the teacher, he would then say: "Pass these [blows] along later to him who has earned them." Nevertheless, with his method, he achieved unusually good results for the entire *Gymnasium* [in mathematics.]

The exact opposite of this man was the religion teacher, whom we feared and hated, for he was malicious and played the spy in many situations. Also, our *Ordinarius*, a classical philologist, garnered little sympathy. I have retained only his nickname "woodpecker". He was kindly disposed to me at first but withdrew his favor in the second semester, because he had differences with cousin Ernst in the *Prima*. My cousin's anger toward this teacher was so great that he allowed himself to get carried away [by pulling] a stupid prank, which might have brought serious detriment to him. Together with another *Primaner*, in the dead of night, he opened several windows in

© The Author(s), under exclusive license to Springer Nature Switzerland AG 2022
D. M. Behrman and E. J. Behrman, *Emil Fischer's "From My Life"*,
Springer Biographies, https://doi.org/10.1007/978-3-031-05156-2_3

the teacher's apartment. I only learned of this deed after it was done, and luckily, I remained the only other one who knew about it. The prank caused a great uproar in the school, and the woodpecker made tempting promises for a tattler, but the secret remained secure.

Aside from school supervision, we lived in Wetzlar similarly to students, for the host of our house cared not at all about our doings. Every night, the key to the house lay in a groove in the plinth under the doorway. When we needed a school absence excuse due to sickness or [for some other reason] he would sign any note we laid before him. We called him the Coffee Cap, due to a round cap that he usually wore. His brother had the nickname "The Polar Bear."

Cousin Ernst and I had a living room and a bedroom together. In addition, there were two other pupils of the *Gymnasium* and one soldier serving a one-year voluntary term quartered in the house. The cost was quite cheap. For apartment, heating, and complete board except alcoholic beverages, the yearly cost was only 110 thalers, thus about 1 mark per person per day. One might imagine that [for this price] the meals would not be sumptuous. On the contrary, the rutabaga and prunes with pork played a leading role. I can still see the Polar Bear by the stream that ran next to our dwelling house, washing the rutabagas designated for our nourishment in a vat. I would never have dreamed that 50 years later the rutabaga would become so widely used and so cursed a source of nourishment for the German people.

Supervision of our lives by the school was not overly strict. [We] visited each other in our rooms, and our house formed a popular assembly point. Unfortunately, most of the other pupils whom I associated with were considerably older than I; for there were many farmer boys among [us] who were determined to study only at maturity. One of these, who still sat in our class, was already 21 years. Naturally, the habits of these young men were different from those appropriate for a boy of 13 years. Their company therefore was not advantageous for me. I was lured to smoke tobacco, drink beer, and play cards, and generally the conversation was crude and indecent. All this certainly had an influence on my health, and I am afraid that the cause of my later stomach illness was immoderation in smoking and drinking at that time.

Luckily, music offered a counterbalance to these unfavorable influences. Cousin Ernst was unusually gifted musically, and for a while even had the intention to become a professional musician. In our eyes he was nearly a virtuoso on the piano, but he also played violin, cello, and flute. Next to him I was a bungler, but through assiduous practice on the piano I was able to play sonatas by Haydn, Beethoven, and Mozart, etc. tolerably well. Cousin Ernst then invited me to play together with him, and as there were several other musical pupils, we frequently played trios and quartets in our room. Our music was so good that many people in the neighborhood took part. Cousin Ernst and I were able to meet school requirements without much trouble.

School attendance was interrupted in the summer of 1866, a welcome interruption, due to the Prussian war with Austria and the southern German states. The Prussian district of Wetzlar was surrounded by enemy states, namely by Hessen-Darmstadt and Nassau. After the Prussian soldiers had marched out, we remained cut off for several weeks due to the interruption in the train- and postal service, and then came a much more important event–the invasion of enemy troops. It was a division from

Baden, if I am not mistaken, which invaded from the vicinity of Giessen. As soon as we had word of this, we [became the subject of] the enemy troops' curiosity, and I and several other pupils enjoyed the pleasure of being taken captive by an outpost. We were properly interrogated and then released again. To our especial pleasure, however, it went much worse for the unloved religion teacher of the *Gymnasium*. He had wanted to look at the foreign soldiers, but they immediately recognized his sneaky nature, which we pupils so feared, and held him for several days. A few days later the enemy marched into the city of Wetzlar and occupied the public buildings, including, to our pleasure, the *Gymnasium*. We had perhaps 8 days unexpected vacation from school and therefore enough leisure time to engage with the enemy troops.

At that time, war was more agreeable than it is today. To be sure, the soldiers were provisioned at the city's cost; however, they behaved very properly, and the relationship between them and citizens of the city and, in particular, us *Gymnasium* pupils, was quite friendly. [Then] they withdrew, because Prussian forces had in the meantime vanquished the main army of the south-German states through powerful strokes at Kissingen and on the Main. The next soldiers who moved into the city were unfortunately all wounded, mostly Prussians and Bavarians, for whom we felt sincere sympathy. Then came also [an] old Prussian militia, which was distributed in the neighboring villages of Hesse and Nassau as so-called *Fresskompagnie* for pacification of the somewhat rebellious population. With the victorious battles in Bohemia the war soon came to an end, and school instruction was resumed with double severity until the vacation.

The autumn vacation of this year was thoroughly spoiled by the confirmation, which took place in Flamersheim; for in four weeks, I had to learn by heart the entire Heidelberg catechism and a succession of hymns, [since] I was previously unable to participate in the confirmation instruction. I succeeded in this, a consequence of my good memory; however, it did not serve to strengthen my faith, which already had mostly been lost by reading the *Life of Jesus* by David Strauss, through the influence of my older school fellows in Wetzlar. Indeed, it was as a free spirit, though with a certain feeling of shame, that I stepped to the table of the Lord, and I have never regained the positive faith that I possessed in early youth. Against that, I have in later years come to the conclusion that religion is an important part of human culture and that the Christian faith can give the individual great moral value and inner happiness. My own mother and many of the Catholic school fellows or student acquaintances were proof of this. Later, in Berlin, I regretted, for my own sons, that the influence of the big city and unfortunately also the atmosphere of the Berlin schools is not very favorable to the cultivation of religious feeling.

A second year at Wetzlar passed without political disturbance. Cousin Ernst finally had his diploma, and I was promoted to the *Prima*. In the meantime, I had had my fill of Wetzlar and therefore left the school with the cousin. Memories I have retained from the stay in Wetzlar [include] the beautiful countryside, the old-fashioned architecture of the old city lying on a mountain slope, and the numerous records of Goethe that we treated with reverence. It goes without saying that *Werthers Leiden* was a much-read book [among us], though we were not influenced by the character's morbid temperament. I have never again come in contact with schoolmates from this time, and I have also not seen the city again.

During the autumn vacation, which I usually spent at home, the next prospective school for me was the Friedrich Wilhelm *Gymnasium* in Cologne. My father made the necessary application to the director, a Professor Jäger, who had become famous as author of a world history, and who had been an excellent teacher at a school in Württemberg and then called to Cologne. The school was overcrowded due to its good management. Nevertheless, the director had reserved a place in the *Prima* for me because he felt obligated to my uncle, the surgeon, whom he credited for saving his wife from mortal danger. But in the letter he sent my father, my acceptance was not clearly stated, and the overcrowding of the school so heavily emphasized, that we had the impression our application was denied.

30 years later I got to know the director at the second school conference [held] at the State Ministry for Education and Culture in Berlin. At that time the misunderstanding over the supposed denial of my application was cleared up, and the jovial old gentleman made the joking remark: "What you could have become, if only you had come under my cane."

Much time had been lost on the proceedings at Cologne. The school year had already begun, and I had to find a place somewhere else as soon as possible. The choice came down to the *Gymnasium* at Bonn, where I was accepted on my 15th birthday. My entry into the school was completed under apparently less-than-agreeable conditions. It was just the hour for mathematics, and when the old teacher, Professor Zirkel, the father of the well-known professor of minerology at Leipzig, saw me, he angrily asked my name and origin and made it clear that my late arrival was inconsiderate, for he would now be obliged to change his list again. Soon thereafter came a knock on the door, and the pupil who had been sent out for correction came back laughing. "The father of the new *Primaner* is outside and invites his son to appear at the *Stern* guesthouse for the midday meal at 1 o'clock." That brought renewed surges of anger from the teacher, and to the exultation of the entire class, he became so vehement in his reproaches, that I had a mind to leave the school and refused to answer his further questions. Then he noticed he had gone too far. He altered his tone, and, in order to reconcile, asked after Euskirchen and the Flamersheim forest, which he knew well. Only now did it become clear to me that he was basically a kindly man, and after he discovered my interest in mathematics, we became best friends. Many of the other *Primaner* were not very deferential to him, and when he became angry with them, he would spend the rest of the hour working problems on the board with me alone.

The *Gymnasium* [in Bonn], in general, as to instruction, was not the equal of the school in Wetzlar. In part this was due to the predominance of the religion teacher, a catholic cleric. He exercised a true tyranny, under which, to be sure, we Protestants did not suffer, but the scholarly instruction certainly was harmed. The second reason was the age and illness of the director. He had previously owned quite a good reputation as philologist and, in particular, as translator of Horace. However, in my time he was no longer capable of work and died soon thereafter. Also, his successor, a small intellect, did not succeed in bringing new blood into the school.

Among the teachers, the most proficient was a Mr. Deiters, who was later called to the provincial school collegium in Coblenz; but we feared him because of his sarcasm and strictness. Aside from the mathematician Zirkel, the German teacher Remacly was the oddest personality. I have also retained a vivid memory of the *Ordinarius*, a classical philologist. He had great difficulty maintaining discipline, due to extreme short-sightedness. He had a hobby of collecting old coins, and one could give him great joy by bringing him Roman coins, which at that time were commonly found in the Rhineland. His tendency to retain all pieces brought to him for inspection led one day to a cheeky practical joke. Two jokers in our class took trouser buttons bearing the name of the Bonn tailor "Hannes" and altered them by physical and chemical means so that they resembled well-weathered Roman coins. These were then delivered to Mr. *Ordinarius* as curious discoveries. He was knowledgeable enough, however, after examination, to determine the hoax and then expressed his justified indignation. He was also greatly interested in Roman inscriptions, and he imparted to us the skill of making reproductions of them on pasteboard. It brought me joy to deliver to this man, whose scholarly efforts filled me with admiration, reproductions of several Roman inscriptions from the area around Zülpich, the supposed site of the [Battle of] Tolbiac of the Merovingian times, and from Weingarten, near Euskirchen.

At the Bonn *Prima*, for the first time, I had religious instruction that I enjoyed, that genuinely interested me. The teacher, who at the same time was a Protestant pastor in Bonn, had us read the New Testament in Greek. He knew how to portray the characters as so lively and in context with the major historical events of that time, that I got a concept, not just of the moral power of the Christian doctrine, but also of the huge intellectual, social, and political movement that it unleashed. From this man's portrayal I became familiar with the old Christian practices, the catacombs of Rome, and similar things before [reading about them] in the original.

Aside from school I got on very well in Bonn. I lived with a family Kemp in the Bonn Lane, only a few steps from the *Gymnasium*, and the food was splendid. Aside from myself, there was the *Oberprimaner* Fischenich, son of a farmer from the area around Flamersheim, good musically and a good comrade. I was happy that my son Hermann became acquainted with the son of this old school comrade on the battlefront in Lorraine and brought him to me in Berlin as a guest about one year ago. In addition, there were two much younger pupils from the lower classes. We all assembled for meals with the family and employees of the business. The host ran a profitable business with his son "Paul Kemp & Son." [They sold] notions, student commodities of all kinds, furs, and even Carnival costumes. This brought heavy traffic into the house and many amusing incidents. I spent two truly cheerful years there. At the same time, my cousin Ernst Fischer was studying medicine in Bonn, and although his room was in another house, nevertheless we sometimes played music together.

Naturally, we also became familiar with the student life in Bonn, both legitimate and unwarranted features. At that time, I enthusiastically took instruction in fencing with saber at the university fencing room. The result was that later at the university I no longer had any interest in these things.

In August 1869 I passed the *Abitur* exam, and the exit certificate indicated that I was not a bad pupil. Nevertheless, I have, to my regret, no friendly memories of the *Gymnasium*. I feel obliged to emphasize that here, because I have the feeling that the humanistic *Gymnasium* does not fulfill the obligation which is assigned to it and does not, [contrary to] what is mostly maintained, give its pupils the general intellectual maturity that is needed for higher education. I am speaking here not as a researcher of nature, who must always complain that [he received] inadequate mathematical instruction in school. My judgment concerns the language instruction, which, in its present form, certainly [places] an overemphasis on knowledge of grammar. How much precious time did we have to spend on absurd memorization of rules! The rarest exceptions of a declination or conjugation, that even career philologists scarcely use in practice, we had to know in order to be a good pupil. Almost never did instruction concern itself with the beauties of classical literature [or] the close connection with the admirable general Greek culture. I am thoroughly convinced that if our teachers had placed the main emphasis on these things, most of us would have followed with enthusiasm. [As it was,] we simply lost the taste for classical antiquity and were glad to give up these studies after the *Abitur*.

To celebrate the final exam, I was able to invite the entire class to my host's garden, for my father had donated a large barrel of beer from the Dortmund brewery.

Chapter 4
University

At the age of 16¾ years, it was now time for me to choose a career. According to my tastes, I would have preferred to become a mathematician and physicist, but my father held these disciplines [to be] too abstract and the possibility too small to assure making a living. He therefore couched his advice in these words: "If you absolutely want to study, then choose chemistry," the practical, useful side [of chemistry] being known to him through his business ventures. Secretly he may well still have had the hope that I would choose a business career, as I was the only son, and he would understandably have been glad to have me as successor in his business. As I was still quite young, and immediate attendance at university held no great attraction for me, he proposed that I work in a business for 1–2 semesters. I agreed to this and therefore arrived in October 1869 at the wood-handling business of my brother-in-law Max Friedrichs in Rheydt. I was to stay with my sister Fanny, his wife.

According to the strict rules of the business, as the most junior apprentice, I was viewed and treated as such. My duties included picking up the mail, sealing envelopes, and such. I had to keep a small business journal, but this was just for practice and had no bearing on the actual business. Now and then I was sent with a small order to customers, who were small[-scale] cabinetmakers, for the most part. I had no [duties] in the lumberyard, but I was allowed to watch the various work: sawing, planing, transport of the logs, sizing the timber, etc. These tasks which, at that time, were still far from perfect and performed mostly by hand, were much more interesting to me than the pedantic office work. But the entirety was so boring to my taste, that after a few weeks, on the advice of a teacher at the high school in Rheydt, I purchased Stöckhardts School of Chemistry with the [necessary] apparatus and set up a tiny laboratory in an empty room of the business. The experiments were naturally highly clumsy, resulted in a stink or dirty and burned fingers, and made others uncomfortable because of the danger of fire. In the evenings I either played piano or [went] to the guest house to drink beer, smoke tobacco and play billiards.

The brother-in-law was very displeased with my accomplishments, declared that I was the worst apprentice he had ever had, once angrily let fly to other members of the family with this remark: "The boy will never amount to anything." He was later

D. M. Behrman and E. J. Behrman, *Emil Fischer's "From My Life"*,
Springer Biographies, https://doi.org/10.1007/978-3-031-05156-2_4

teased due to this remark, but in general my father also came to the conclusion that a business career was not right for me. He expressed this opinion rather drastically: "The boy is too stupid to be a businessman, he should study."

In the meantime, I had taken a few hours of private study with the previously-mentioned teacher of chemistry and had acquired a cursory knowledge of atomic theory. I can't say, though, that I was deeply moved by this. In the form it was presented to me and as I also learned it from a short chemistry textbook, it seemed feeble and uncertain in comparison with the well-rounded discipline of physics.

In the spring of 1870, I had the misfortune of contracting a stomach illness, probably [resulting] from a cold. The illness became chronic after I let it linger and failed to seek medical help. Groundwork for this illness perhaps had been laid through foolish smoking and beer drinking [habits] I had started in Wetzlar and continued to practice since then. The acute outbreak of illness made my departure from Rheydt easier, and I returned to Euskirchen to recuperate. This chronic stomach illness was something unknown in the family, and even the doctors at the time understood very little about how to treat it. With the fine methods available today I would have regained my full health in 4–6 weeks. But despite the advice of my Cologne uncle, an otherwise excellent doctor, I had to struggle with this disease for nearly two years, and my digestive tract has never been as strong as it previously was.

I spent the summer of 1870 in Euskirchen, occupied myself in the garden and, to the extent possible, in hunting. On the advice of doctors in the middle of July, on the day of the Prussian mobilization, I went with my mother to Bad Ems to drink the alkaline waters as a cure for my stomach. There was an understandable unease in the Rhineland at the time since it was felt that France had long been preparing for war and would immediately send troops into the Prussian Rhineland. My father had already prepared for it, having brought my unmarried sisters as well as money, securities, and other easily-transportable objects to the right bank of the Rhine. On our journey to Ems, it struck my mother and me as even more curious that we observed absolutely no military preparations as we passed the fort of Coblenz. The only soldier at the fort whom we saw from the train was an officer's orderly pushing a baby carriage. But the Prussian mobilization was already fully underway behind the scenes, and eight days later there were soldiers everywhere. Then came the massive troop transports, of which we in Ems saw quite a bit, since it lay on a major railroad line from East to West. At the beginning of August, the victorious battles at Weissenburg, Woerth, and Saarbrücken took place, and in Germany one had the feeling that the danger of an enemy invasion had been eliminated. Two days before my arrival at Ems, King Wilhelm had departed from there. Recollections of [the king] as well as negotiations with the French ambassador were still much discussed at the spa. Although the number of guests had greatly diminished, the cure took its [full] undisturbed course.

After 4 weeks we were able to return home without the slightest trouble, for the railroad traffic had returned to normal. I was still too young for active participation in the war and, at that time, not healthy enough, for the cure at Ems had not been successful.

At the neighboring house the three oldest sons were already eligible for service and had been called up. Of them, however, only cousin Lorenz gathered military [honors.] As a volunteer for one year [service] in a hunter battalion he took part in the Battle of Metz. During the siege of the fort, he contracted dysentery, returned to Germany, but soon went back—though not yet reinstated—voluntarily into the field, where he participated in General Werder's struggle against the Bourbaki army at Dijon. He was [promoted] to officer in the field. But the military life, which fit his inclinations so well, later brought him bad luck, for it contributed to his inclination to pay homage to Bacchus, which elevated to excess, and he died at the age of about 35 years.

Cousin Heinrich, a trained artillery [man] was more careful; he did not go to war but remained in a reserve battalion in Coblenz.

Cousin Ernst, who had in the meantime passed his medical exam, voluntarily joined the army as an under-doctor. In this capacity, as he later liked to relate, he spent [the war] mostly in the surroundings of Orleans very enjoyably and with little stress.

In short, the war 70 was to all accounts of its participants entirely different from the current one. Losses on the German side were vanishingly small; not more than 28,000 men on our side fell. Also, prosecution of the war was much more humane. A part of the troops were on good, nearly friendly terms with the French population. Also, the war came quickly to an end, and political unification into the German empire came as a most satisfactory reward for Germany.

I was at Euskirchen for the victory at Sedan. I heard the news on returning from hunting partridge. It was said at the time that the war was now at an end. However, this was mistaken, since the Republic under Gambetta stubbornly sought to prolong it. But the French resistance was broken. For us on the left bank of the Rhine, the war persisted only with troop movements and quartering until the end of hostilities.

In the meantime, my health condition had hardly changed. My uncle in Cologne therefore invited me to come in November and live with him for some time, so that he might observe me. His lovely wife, aunt Mathilde, distinguished by her kindness, tended to her ill nephew in the most obliging manner. I got specially-prepared foods and, on the advice of my uncle, as much good Bordeaux wine as I believed I could manage. Also beneficial was the steady temperature of the house, which already had central heating, at that time an unusual luxury. So, I remained the entire winter in Cologne, only returning home for eight days for the Christmas celebration. In my later years I have always remembered this time [with] the uncle and aunt with gratitude, for my condition improved there.

In Cologne there was also greater intellectual stimulation than in Euskirchen. I attended theater and concerts. Above all I was given the opportunity to take lessons in English from an Englishman. As I had nothing else to do, it was not difficult, after 2½ months, to advance to the point where I could write short English essays and orally express myself tolerably. Unfortunately, I did not continue these exercises and was never able to make an extended stay in England, so I never mastered this important language.

Naturally, I became closer to my uncle's children while in Cologne. The eldest, Marie, was already married to Mr. Eugen Coupienne in Mülheim-on-the-Ruhr, but she often came with husband and child to visit. Mr. Coupienne sometimes brought along [captured] French officers as guests in my uncle's house. [These men] were allowed to move about freely in Cologne. It gave my uncle great joy to be able to use the French language again, which he had learned quite well in Paris.

The second cousin Helene, a beautiful, well-nourished girl, who had the nickname "Fatty," was well-known to me from several visits to Euskirchen. She soon followed her sister and married a tobacco factory owner, Carl von Eicken, also from Mülheim. She still lives very happily together with her husband in Hamburg and is the matriarch of 4 prospering children and a great number of grandchildren. Her son Karl, now full professor of ear-and-nose healing in Giessen, was often a guest in my house in Berlin when he was a student, and I have the impression that he inherited the qualities of a good doctor from his grandfather Fischer. The daughter Helene lives with her husband Dr. Kroehnke and 4 promising sons in Zehlendorf and is distinguished by her charm and kindness. A second daughter of the von Eicken couple is married in Hamburg. The second son, a great sportsman, leads the important tobacco company with his father. He is married to the only daughter of my cousin Julius.

The third Cologne cousin, named "Tönn," was clever and inquisitive and received the best grades in school. Later she married a manufacturer Mr. Vorster from Mülheim-on-the-Ruhr and became mother-in-law to the well-known painter Petersen in Düsseldorf.

The only son Fritz, who was called only "the Fischer" by his sisters and was much teased by them, was 4 years younger than I and therefore still at the *Gymnasium*. He was the darling of the father and stayed mostly in his study. It is easily understandable that he followed in his father's footsteps, studied medicine, indeed in Strassburg, where I again met with him frequently. He became associate professor of surgery at the university there and died comparatively young at age 50 years. He was married to Anni née Stinnes from Mülheim-on-the-Ruhr and left one son, who again is studying medicine and, at the moment, is under-doctor in the field. In recent years he became close to my dear son Alfred, who experienced great kindness from his aunt in Strassburg.

During my stay in Cologne, my uncle gave his children strict instructions not to anger the ill cousin and where possible, not to contradict him, even if he make bold assertions from conceited erudition. As difficult as that became for the glib [female] cousins, they nevertheless tried their best to make my stay in Cologne pleasant. As a former *Gymnasiast*, I was naturally on firm ground with cousin [Fritz.] The winter stay in Cologne therefore ranks among my fondest memories. My entire life I have felt obliged, with warm thanks, to my excellent uncle, who has, for me, always remained a paragon, not only as a doctor, but also as a distinguished personality and ingenious person, together with my kind-hearted aunt.

In the spring of 1871, my health had been restored enough that I could go to the university to study chemistry. Living in a guest house certainly was not yet wholesome, but I was welcomed back to my old *Gymnasium* quarters. In the meantime, there had been several changes in personnel. The old couple Kemp had retired and transferred the business to the son, who had in the meantime married the young lady Marie who had been employed in the business as saleslady. The young, obliging woman declared herself ready to accommodate my nutritional needs, and this in fact occurred. Thus, I continued to live as I had with my uncle's family. The host, who had contracted malaria on an earlier visit to the United States and had upset his stomach with too much quinine, learned the appropriate diet for this condition from me. In this way he was also restored to good health and later repeatedly thanked me for the good advice.

In the summer semester 71 I attended lectures in physics, botany, and only a little chemistry, which were not too strenuous but rather quite comfortable. The course in botany [given by] Professor Hanstein took place [starting at] 7 o'clock in the morning at the Poppelsdorf chateau. It was very popular and tailored to the needs of medical students. Taxonomy and anatomy were emphasized while physiology, which would have been much more interesting to me, played only a limited role.

Physics, given by the famous thermodynamicist Clausius, was very boring and all-too lacking in experimental connections. My old preference for physics was thereby somewhat subdued. Only later in Strassburg, when I took the brilliant lecture and good practical [courses] with Kundt, did it again become clear to me what a marvelous discipline [physics] is.

The chemist August Kekulé was the exact opposite of Clausius. He was an excellent speaker, a good experimentalist, and an impressive personality. In the summer semester he gave only a two-hour course in organic chemistry, but the following winter I took his entire course in experimental inorganic chemistry.

During the autumn vacation 71 I visited Blankenberghe bei Ostende, the sea spa, together with cousin Lorenz, who had returned from France and was full of experiences of the campaign. My health was once again good enough that in the winter 71/72 I could begin study in earnest.

Along with lectures now came practical exercise in the laboratory. The institute was in Poppelsdorf, erected in 1864/65 by Hofmann, and it made a splendid impression from outside. Internally the setup was much less beautiful, for example, the partitioning of the rooms and the lighting. In particular, the ventilation left much to be desired. The director gave the large lectures and led the practical work in the organic division. Analytical instruction was entrusted to Associate Professor Engelbach. Without any preparation at all in the performance of experiments one was immediately assigned to carry out a qualitative analysis according to the Will diagram. I was not capable of this and was at a disadvantage in relation to the other [students]. They were mostly apothecaries and generally completed the assignment by illegitimate means. Meanwhile, without any instruction, I made many mistakes and then had to repeat [the experiment.] In addition, the lectures in analytical chemistry, which I was required to take, were very dry and boring. My entry into this discipline therefore occurred under rather unfavorable circumstances, which were very oppressive to me.

The following summer, when I took quantitative analysis, things became even worse. Engelbach had died in the meantime, and Th. Zinke, who had been first assistant and lecturer in the organic division, was his successor. He was certainly a talented and dedicated chemist, but instruction in analytical chemistry was not his area of interest. Although he took pains and sometimes gave me some good advice, it was difficult for him to remove the obstacles that stood in the way of carrying out a good quantitative analysis at the Bonn institute. The inaccuracy of the balances alone was great enough to make it impossible to stay within the usual error bars of the analysis. Use of the water vacuum pump, which had been invented long ago by Bunsen, received an unfavorable judgment in Bonn. One still had, therefore, to use the old primitive [technique of] washing the precipitate on an ordinary filter. This was a difficult test of patience. After I had washed a precipitate of aluminum- and iron hydroxide, prepared, to be sure, without taking the proper precautions, for 8 days without being able to displace the mother liquor, I despaired, so that I wanted to give up chemistry and return to physics. That this did not happen is due primarily to the influence of my cousin Ernst, who advised me to try again at a different institute.

Meanwhile, cousin Otto had also finished with school and had likewise begun studying chemistry in Berlin, and had also then come to Bonn. He did not take the difficulties as tragically as I; nevertheless, he also had a mind to transfer, not just for the sake of the studies, but also to see a bit more of the world. And so it happened that at the end of the summer semester we left Bonn and then, in part due to chance, came to Strassburg-in-Alsace. This was good fortune for both of us, for we thereby came in contact with Adolf Baeyer. I can't imagine anything better that could have happened for me to learn the art of chemical experimentation. I have tried to portray the stay in the Strassburg laboratory in a short essay that Baeyer included in his autobiography, and it has also appeared in print in the introduction of Baeyer's collected papers. I have only a little to add concerning our lifestyle and circle of friends.

In the first semester I lived together with cousin Otto in a very old house [owned by] two very old women, just as the entire city at that time made a truly old and dilapidated impression. This [impression] was made sharper by the massive destruction resulting from the occupation. The so-called Stone Quarter lay for the most part in ruins.

My stomach illness was now cured, but it had left behind a sensitivity of the organ that imposed a certain caution of the diet. The good Strassburg cuisine and the even better French wine, which was available in great supply, were beneficial to me. In the winter semester I [went to] a spa in Baden and, by arrangement with the very considerate host, consumed a large cask of excellent burgundy. This resulted in [weight gain,] so that I was distinguished by my professors from my cousin with the name "the fat Fischer."

My next visit home during the Easter vacation brought the mother great joy, but the experienced father greeted me with these words: "You have got a somewhat drunken head." As a result, I gave up drinking burgundy and turned to the good but much more harmless Alsatian local wine.

The majority of [my] colleagues, many of whom were from the Rhineland, visited the wine houses. There was also beer of various quality: locally brewed in Schiltigheim, as well as Bavarian, but we visited these houses only in the summer, [when] it is unusually hot in Strassburg and therefore stimulates the consumption of beer.

Student life, which is customary at other German institutions of higher education, was [only] slightly developed at our time. There were several fraternities, but they played no great role, because there was no sounding board for them among the population, and the entire tradition in the city was lacking. Against that, the casual intercourse among students in the guest houses and also in the lectures was very pleasant. The many foreign elements, Russians, Poles, and other foreigners did not disturb this. The Alsatians held back at first but came to the institute in later semesters. Among them were several well-educated young men with very pleasant manners. We did not come in contact with the cultured families of Strassburg, but I did get to know many ordinary people, particularly on several outings to the Vosges, and have retained a very good impression. With the exception of Südelsass, where the textile industry was well-developed and their business interests were directed toward the rest of France, the population throughout displayed their old German extraction. The people spoke their Alsatian German with French crumbs mixed in, but for the most part they were not capable of speaking French. In fact, they were nothing less than German-friendly, and they liked to complain about the "Schwoben." But when politics were set aside one could get along with them quite well. It is a mystery to me why the German administration has not managed in 40 years to earn the trust of this basically good-natured, indeed democratically-minded people.

In Strassburg itself many remnants of French installations and customs still remained. In public dance halls many still amused themselves with the cancan, and it was comical to watch individual farmer boys at a Sunday dance attempting the usual cancan jumps, though seldom with success. Also, the manner of respectable discourse between the sexes was too free for the German notion.

Discourse between students and professors, of whom nearly all were still at a young age, was more comfortable and more engaging than at any other institution of higher education. In particular, the science researchers and medical students benefitted. The institute was richly provided with money by the German government, and the number of students remained limited in the first years, so the practical instruction was excellent, which also soon showed in the scientific achievements. This advantage of Strassburg gradually increased its attraction, and if the city had not been decried as so expensive, the number of students would soon have equaled those at the old German universities.

Our descriptions of academic life at the young university in the old imperial city also served as publicity within the family. After a few semesters, cousin Ernst came here after he had passed the state medical exam and had been employed for a time as a junior doctor in the surgical division of the Cologne City Hospital. In Strassburg not only did he visit the surgical clinic, which was under the direction of Professor Luecke, but he also took interest in studies in anatomy with Waldeyer and in pathological anatomy with von Recklinghausen. He also took the opportunity to make a small discovery, with which his name likely will be permanently connected, and I want to mention this because I was allowed to give the suggestion. At that time, the dyeing of anatomical specimens was generally quite common. [The technique] was introduced in the science by my father-in-law Josef von Gerlach with the application of carmine red. However, the number of dyes used was limited. When I heard this from cousin Ernst, I advised him to try eosin, which had recently been discovered by Baeyer and Caro. I had become familiar with it in my doctoral work on fluorescein, not only as a chemical preparation, but also on my hands as a splendid dye for animal tissue. The cousin's experiments also turned out so well that he could introduce the dye into anatomical practice, where it remains today.

[Cousin Ernst] was particularly interested in antisepsis in surgical practice, which had acquired great practical importance with the Lister [surgical] dressing. He attempted to improve this and was already of the opinion that antisepsis in surgery would follow asepsis. Next, he attempted to find less harmful antiseptics than carbolic acid. He arrived at naphthalene, whose toxicity for vermin was well-known. He publicized the use of this hydrocarbon and recommended that it be used in France against Phylloxera (wine louse), however without success.

There was another invention connected with these efforts, whose [discovery] is reminiscent of an anecdote. Following Ernst Fischer's recommendation of naphthalene as an antiseptic, two lower-doctors of the Strassburg clinic for internal medicine had the idea to use this hydrocarbon for disinfecting the diseased bowel. They ordered the remedy from a Strassburg drug handler, with the remark that it must be completely pure, and they received a preparation, which the business claimed was so pure that it no longer had an odor. When they gave this remedy to feverish patients with bowel disease, they observed a rapid decrease in temperature. Luckily, one of the doctors, a brother of the chemist Eduard Hepp, was also well-trained in chemistry, and he soon came to the conclusion that the remedy [they] used was not naphthalene at all. The chemical investigation by Eduard Hepp showed that the [remedy] was acetanilide, which the druggist had delivered as naphthalene after apparently switching the bottles. Thus, acetanilide, under the name of "Antifebrin" became a well-known remedy for fever, still used extensively in east Asia. In Europe it has been replaced by phenacetin, which is essentially a modified Antifebrin; it has the same [active] atom group.

It seems to me useful to depict these connections, not only on historical grounds, but also as a fresh example that chance frequently plays a role in invention.

Cousin Ernst later turned exclusively to surgery, became lecturer and associate professor at the university, and owned a private surgical clinic in the Cooper Lane, previously described. He died during the war of typhus and a lung inflammation. Soon thereafter his only son died on the Western Front. Two of his daughters are married in France. Ernst converted to Catholicism out of love for his wife and children. I don't believe that [he became] religiously convinced; for I have always known him only as a perfect free spirit.

Somewhat later yet another cousin, Fritz Fischer, son of the surgeon in Cologne, came to Strassburg as a student of medicine. Just after beginning his studies, he contracted scarlet fever, probably the result of an infection from anatomy [lab.] As already mentioned, he remained [in Strassburg] and also died as associate professor of surgery.

Among the other student acquaintances, I must mention Joseph von Mering, with whom I later carried out several chemical/medical works. At that time, he was quite a character, a true son of Cologne, a humorist, possessed of unusual bodily strength, and feared as a dangerous saber fencer.

The summer months in Strassburg were unusually hot. The swampy areas near the Rhine and Ill brought an oppressive plague of mosquitoes. The frogs also thrived in the swamp, and these [in turn] were doubtless prerequisite for the numerous storks that made their nests on the high gables of the old houses. It was an amusing sight, as one stood on the platform of the Strassburg cathedral, to look down at the city and catch a glimpse of the hundreds of stork nests. The warm summer and the position of the city, [situated as it is] between the Black Forest and the Vosges, brought unusually severe thunderstorms. From my apartment in the Calf Lane, quite close to the cathedral, I often admired the lightning as it played from the lightning rod on the cathedral to the storm cloud. On such hot days we liked to drive to Kehl, to bathe in the Rhine and later to drink the local wine of Baden in a good guest house. We chemists also arranged a farewell party here for Professor Baeyer, at which I gave my first after-dinner speech. I got stuck, which aroused great exultation; this [served] as a reminder to me to be better prepared at similar opportunities in the future. By the way, bathing in the Rhine was not a perfectly safe activity; for the current in the middle of the river amounted to 3 m per second, and the push of the heavier pebbles was so strong that one could easily observe it in the water.

In the summer the beautiful surroundings of Strassburg seductively enticed [us] for outings. Mostly we went to Baden. The Black Forest and the Ode Forest became familiar to me early on. Twice I also beat a path to Switzerland. One journey, that I undertook with cousin Julius, has already been described. One year earlier I took a merry tour through the Bern highlands with cousin Otto Fischer and a student of pharmacy from Bavaria. The end of this round-trip in the southwestern corner of Germany consisted [of] a visit to the southern Vosges in the company of cousin Ernst Fischer. We marched up to the crest of the mountain that forms the border between Germany and France and finally came in the canyon to Belchen, where there is a very good guest house of the same name. A large company of pure French was assembled here who had made the trip across the border. We remained isolated at table, our conversation with the neighbors limited to a few polite phrases. But suddenly a woman in the company took ill with abdominal bleeding, and a doctor was urgently needed. As there was no one else [suitable] in the company, cousin Ernst volunteered his help and soon eliminated the danger. From that point on the tone of the company toward us was completely changed. We were invited to take part in games, and they outdid one another in their pleasantries toward us. However, we soon had to depart, in order to reach our goal, the city of Mülhausen, the same day.

The next day I separated from cousin Ernst, who wanted to get to know the hospitals in Mülhausen, and drove alone back to Strassburg. On this occasion I became acquainted with Otto N. Witt in comical fashion. He was sitting across from me in the train and asked after a short time: "Sir, you are surely a chemist?" I affirmed this and laughingly guessed that he [deduced] this from my strongly-colored hands, adding that he seemed not too far removed from such work himself. The diazo-group, with which we both had previously worked, was the culprit and brought us into a collegial connection. We introduced ourselves and related our current work to each other. I was occupied with phenylhydrazine and Witt had at that time just discovered chrysoidine. He was coming from Zürich and going to England, in order to make practical use of his discovery. I invited him to stop over in Strassburg, whereupon he agreed, and we then spent a very enjoyable day together. Many years later, when I came to Berlin, it was my pleasure to renew this acquaintanceship. The closest we came was when we both took the cure at Kissingen in the von Dapper sanatorium. There we had adequate leisure time to engage in casual conversation, not just scientific matters. Witt was a performer-type, very well-read, knew several languages, had a great sense of humor and an impulsive character. One only needed to be very careful [not] to offend his wife. His wife at that time was a very pretty and charming English woman.

Fig. 5.1 Munich Times

Chapter 5
Munich

Several weeks after the trip to the Vosges, I left Strassburg with cousin Ernst to make a trip to southern Germany. We got to know Tübingen, Stuttgart, and then Munich and had many enjoyable experiences. In Tübingen patriarchal conditions reigned at that time. When we turned into a wine house, the host's little daughter immediately sat before us to give us information about the city and goings-on in the city. The director of the chemical institute, Professor R. Fittig, had already been called to Strassburg as Baeyer's successor. When we asked for a [guide] to show us the institute, a young lady appeared, who was employed in Fittig's house. She knew the institute as well as a career chemist and made a charming [guide] (Fig. 5.1).

In Stuttgart there was naturally far more to see, and in the Bavarian capital we met with Professor Baeyer, who had already moved [here] several weeks earlier. It was fun for him in the evenings to lead us to several amusing taverns, nominally to make me like the city. I still remember that one of the bars was in the neighborhood of the *Hofbräuhaus* and had the classical name "Orlando di Lasso."

Of course, we thoroughly examined the art treasures of Munich, but I must openly confess that in contrast to the great impression that I had of the pictures of the Pinakothek [museum], the remarkable sculptures of the Glyptothek [museum] left me relatively cold. Also, for most people the enjoyment of art must be cultivated. The meaning of antiquity for sculpture and architecture became clear to me only in later visits to Italy.

What interested me most in Munich at that time was the scientific and material life; for it was a question of whether I would follow my teacher and tie myself down there for an extended time. My parents, particularly my mother, had decided against it, because Munich had quite a bad reputation at that time for reasons of health. Typhus was so widespread that young people from the Rhineland, if they had not been immunized by previous illness, could be nearly certain of getting infected in Munich. And two years earlier, in the year 1873, cholera had also [caused] devastation. In addition, the climate of the high-altitude city with [its] rapidly changing temperatures was feared in the Rhineland.

© The Author(s), under exclusive license to Springer Nature Switzerland AG 2022
D. M. Behrman and E. J. Behrman, *Emil Fischer's "From My Life"*,
Springer Biographies, https://doi.org/10.1007/978-3-031-05156-2_5

Professor Baeyer set my [mind] at ease over these matters and promised me the hygienic protection of the Pettenkofer School. This was also later the case extensively. Through Baeyer, we met the Pettenkofer under-doctors, particularly Dr. Forster, a very charming and knowledgeable man, who later was Professor of Hygiene at Amsterdam and finally at Strassburg-in-Alsace. He had a typhus chart of the city that showed not just the suspicious streets, but the individual houses marked for typhus infection. We used this chart in selecting our quarters in the healthy area near the institute and always followed Forster's advice on hygiene. When I left Munich for further travels at that time, I had decided to return here in late autumn to continue my studies of hydrazine compounds in Baeyer's laboratory.

I went next via Salzburg to Vienna, whose advantages the Austrian students in Strassburg had related so well. In fact, my expectations were not disappointed. The splendid setting in the vicinity of the mighty river at the foot of the beautiful, forested mountains, the old city rich in historical memories and the splendid new construction on the Ring [Road,] the great art treasures and the excellent theater, the charming innocent manner of the people, the many beautiful women and the good, natural food were indeed likely to captivate a young person such as myself, and I would perhaps have traded Munich for Vienna, had not the visit to the chemical institute disenchanted me.

It was indeed a new building in an extravagant style, but it did not at all give an impression of practicality. The people suited me much better. They were busy despite [the institute] being on break. I next met with an old acquaintance, a Strassburg student colleague, Dr. O. Ziedler, who introduced me to the other assistants. Among these Dr. Zdenko Skraup stood out, and I immediately made friends with him.

At the midday meal I also met Professor Weidel, who was perhaps 10–15 years older than we and seemed somewhat reserved in conversation with young people. I saw Professor A. Lieben, who led one division of the institute, only from a distance. Only later did I come in closer contact with this refined and charming colleague.

Skraup and Zeidler were the leaders in that part of the Vienna nightlife that is not described in the travel books. We also took two small outings in the vicinity of Vienna, and I attended the grand opera with Skraup. His [finances] forced him to live modestly, and we therefore [chose] the cheap standing area, which was perfectly fine with me. Officers were also allowed there, and they were admitted for half-price. As Skraup was a reserve officer, he immediately announced on the grounds of thrift he would wear his uniform to the theater. But for this he had to shave. As the time was very short, I declared myself ready to perform the task. Zeidler lent his straight razor, which we later learned was used for many other purposes. When I was only half-way through the operation, Skraup declared that he would rather have a tooth extracted than continue to abide this torture. He sprang up and ran, half-shaved, still covered in soap, to the nearest shaving salon, which was some distance away. Since then, I have never again performed an operation on another person's body.

After a most enjoyable stay of perhaps eight days, it was time to leave Vienna, where I later twice returned for scientific research meetings. The return journey proceeded via Nürnberg, Würzburg, and Frankfurt-on-the-Main to Euskirchen, where for several weeks I could assist my father hunting woodcock and hare. At the same time, I sought to make clear to my parents that I had made a handsome discovery with phenylhydrazine and that it was my wish to pursue this [research] further as a free scholar at the Munich laboratory. Although my father maintained that he did not understand the matter and questioned the usefulness of the discovery, he left the decision entirely up to me. He had already made me self-sufficient, for as soon as I had reached the age of 21 years, he gave me the same sum as my sisters had received for dowry. The interest sufficed for my needs, which were not extravagant. I gladly left the capital with my father, who continued to manage this [money] until I was 40 years old and moved to Berlin. At that point he declared that I was old enough to take care of such things myself. My mother's fears concerning the health hazards were assuaged, and so, in the middle of October, I moved to Munich.

The laboratory that Baeyer had temporarily established had been built according to plans by Liebig 20 years earlier. In his appointment at Munich, the great researcher had explicitly declined the responsibility of providing practical instruction for students. Therefore, he held only large lectures in chemistry and devoted himself in the laboratory exclusively to his own research. By the way, [his] time of great experimental investigations had already passed, and the Munich time was devoted more to work in the literature. Consequently, the laboratory, which included Liebig's apartment, was quite small. On Baeyer's arrival the official apartment was sacrificed, and a provisional laboratory was established. At the same time preliminary work was begun on a new building.

[Baeyer's] assistants in Strassburg, Dr. E. Hepp and Dr. C. Schraube, as well as the superintendent Kamps and the servant Gimmig had moved [to Munich.] We met the only previous [faculty] member of the institute, Associate Professor Jakob Volhard. For many years he had represented his uncle Liebig in lectures and also provided assistance in editing the *Annalen der Chemie*. He was a clever, refined man and an attractive personality. He was distinguished by his unusual stature and physical beauty, urbane, and inclined toward humor. He had acquired, through interaction with entertainers, a brisk tone free of all pedantry. The open manner with which he expressed himself, which engendered trust among all and was [agreeable] to us young people, could, however, rise to extreme crudeness. This happened when he had a violent fit of temper, of which I myself experienced a test. One day he encountered me during an elemental analysis and began a pleasant conversation, when his glance fell on a pipet that had been given me by a servant as a worthless piece and had been used for a common purpose. In actuality, it belonged to a set of pipets that Volhard had personally calibrated for analytical purposes. And now there followed so sudden and frightful an outburst of anger, that I feared it would come to blows, and I placed myself in defensive position. Luckily, the fit was soon over. He was sorry for the violent temper, and we were easily reconciled and have remained friends ever since. Twice I have even become his successor, in Munich and also in Erlangen.

The lectures, that Professor Baeyer punctually began in Munich, enjoyed quite good attendance from the outset. In the small analytical division of the institute, whose direction Professor Volhard had taken over, considerable activity also prevailed. Against that, in the first semester, visits to the organic division remained modest. Aside from myself there were perhaps 6 *Praktikanten*. I can't remember

[all] the individuals, but if I am not mistaken, Paul Friedländer and Wilhelm Königs were among them. The latter announced his quirky personality immediately, for he arrived at the institute wearing one shoe and one slipper and also remained this way for a week, although seemingly without any pressing need. People who care so little about astonishing other people have always appealed to me, and I therefore gladly sought the acquaintance of Königs, and it happily turned out that Königs, being from Cologne, shared his region of origin with me. His outer mantle was not beautiful, but it contained a beautiful soul. I have not met many people in life who, with such eminent understanding and such a gift for jokes have such a genteel mindset and so charitable a judgment for the error of their fellow humans. He has this characteristic to thank for [his] general popularity, and I myself have always been glad to count him among my friends.

Theodor Curtius dedicated an engaging biography to him, but that can't stop me from remembering him here and portraying several experiences. Königs was open-minded, had a complete understanding for scientific things and was not lacking in ideas for experiments. Against this, his practical skill was [somewhat] lacking. As he often lamented, it was not possible for him to carry out or even watch over different trials at the same time. With this came a certain sluggishness in the practical work, which in part was due to his earlier chemical training. He came from Bonn, where he also had earned his Ph.D., and where rapid work was apparently not cultivated. He related on his arrival that it usually took him 2–3 days to complete an elemental analysis in Bonn. When I laughingly replied that I could complete 5 in one day, he would not believe it until I demonstrated it for him. To be sure, double apparatus is necessary for this. Without the sluggishness in experimental [work,] Königs with his talent and wealth of ideas in science would have achieved much greater things.

In the company of young people, Königs was always the center of attention due to his quick wit. He also had the gift of producing quite pretty poems for special occasions and small plays. Several of these would be worthy to appear in print. They would doubtless comprise, similar to the works of the Berlin chemist Jacobsen, a worthy enrichment of the humorous chemical literature.

The content of one poem, though unfortunately not the wording, remains in my memory. He composed it for a small celebration of the chemical colleagues in Munich on the occasion of my departure. It was modeled on Scheffel's Guano Song and pertained to my discovery of the preparation of caffeine from xanthine and guanine. The information came to the birds on the Guano Coast through a booklet of reports. In order to defy the German competition, an old bird makes the suggestion to younger colleagues: "From now on, shit in homologous rows" and "when the homologous product is finally achieved, a special sample will be pressed out to honor Mr. Fischer." The poet then added to this little song yet another verse of ridicule: "So sang 2 lively birds at the expense of the President and drank from the wine that he never mentions to others." This had to do with the following connection: at that time, Königs and I had the same landlady and lived on the same floor. We often visited each other late at night in order to drink a glass of wine together. Shortly before the celebration I met Königs on returning home. He was in his room in the company of a young chemist, both apparently somewhat dismayed by my entrance. Königs had just composed the

Guano Song and had helped himself to one of the best wines in my wine cellar. This had hastily been replaced with a coachman's wine on my arrival. The young colleague was a conspirator, for when composing poetry Königs needed wine and company, and he let such assistants keep the meter, so as not to err. After I left, he added to the poem the above-mentioned verse.

My wine supply at the time was by no means small, and several [bottles] were quite good. On my move to Erlangen at Easter 1882, I gave the landlady the task to send my wine. The wine never arrived, and an inquiry revealed that Königs had drunk it with his friends. Instead, a large barrel of Munich beer appeared as replacement for the missing wine. He knew well that I would just laugh over the joke as he would have done in the same situation.

I repeatedly traveled with Königs, the last time May 1902 to Nervi on the [Italian] Riviera. The third in our group was S. Gabriel, likewise quite a joker. When we went out together in the evenings to sit down with a glass of beer, the two colleagues occupied themselves by outdoing each other with jokes. It was like rocket fire, and for an hour I did nothing but laugh. Even more comical than the jokes was the unusual pleasure the two gentlemen took in their side-by-side [repartée.] Those were cheerful days that we spent taking splendid walks in the lovely countryside, [and] amusing things happened here as well. One day we turned back to the city on a very steep, paved footpath and met three Italian boys of approximately 10–12 years, who immediately made remarks about the *forestiere*, without suspecting that we understood something of the language. "*Che banda senile*" remarked the first, "he will surely fall down," declared the second and then immediately made a bet which of the three would meet this fate. It was only a few seconds until Gabriel literally was sitting on the ground, which the young band naturally found extremely amusing. The old band had, however, fundamentally taken part in the joke.

Another journey to Italy, that I took with Königs from Munich, led to Naples. Due to his red hair and his bifocals, he was recognized as a foreigner at a distance by cab drivers, beggars, and similar types and correspondingly implored. This sometimes led to comical situations. At that time, the cab drivers of Naples had the habit, not only to invite foreigners to use their vehicles, but even to follow them on the street and literally drive into their path when they wished to cross the street. Our friend Königs recognized the opportunity to reply to this importunity with a tease of his own. He climbed softly into the vehicle and then quickly out again on the other side. When the cabbie drove off, the vehicle was empty. However, scarcely had he carried out this experiment a few times when the cabbies [devised] a counterstroke. Namely, as soon as he set foot in the vehicle, it was set in motion. Then great laughter ensued and Königs had to pay.

The main purpose of the journey, however, was never lost on these stupidities. Königs was refined enough to appreciate the enormous art treasures of Italy and the sites of antique culture. He could get enthusiastic about good paintings. The example of his only brother may well have contributed to this. His brother was a banker in Berlin and used his substantial income for the purchase of good paintings. Upon his early death, the greatest part of the valuable collection was donated by the siblings to the national gallery.

In later years Königs visited me regularly, at least once a year, in Berlin, and then sometimes brought me in contact with his siblings, particularly with his clever and refined sister Miss Elise Königs, who is well-known to scholars in Berlin as a female patron of the arts and who was honored by the Academy of Sciences with the golden Leibniz medal.

Paul Friedländer, the son of a university professor in Königsberg, was completely different in character from Königs. My memories of Munich are closely tied to his personality. He arrived in Baeyer's laboratory as a student; however, he was so gifted that he was seen as an equal by us older chemists. He was also very musical and could play quite difficult classical pieces on the piano from memory. He made his name in the science during his later work on thio-indigo and antique purple, which he recognized as dibromo-indigo. I also took several trips to Italy with him, [including] my first, which [led] to Verona, Venice, Padua, Florence, and Milan. He knew the language better than I and therefore was a valuable guide. He is 6 years younger than I, still in good health, and active at the technical college in Darmstadt. Due to the war effort, which he took part in at the suggestion of Professor Haber, I have come in contact with him more often.

In the spring 1876 the cousin Otto Fischer returned to Baeyer's laboratory after he had worked on methylanthracene with Liebermann [during] the winter semester. Soon thereafter we both began a collaborative investigation into rosaniline, to be discussed later.

At roughly the same time our circle was joined by an older student of chemistry, Hans Andreae, from Dresden, whose mother was born Dilthey, from Rheydt, and therefore was my cousin. The father, an artist painter, stemmed from the Rhenish family Andreae, which had a large, very well-known velvet factory in Mülheim-on-the-Rhine. This Hans was a fresh, cheerful fellow, but very lazy and given to [drinking] Gambrinus. In Munich he worked hardly at all, but he did drink plenty of beer, played cards, and pulled pranks. His foolish lifestyle [led] to not infrequent accidents. He had already suffered a stab wound in the lung at a brawl in Leipzig. In Munich he suffered from a protracted dislocated foot bone, and one semester later he fell ill with severe typhus. After this illness he returned to his old lifestyle, [whereupon] I advised his father, who had turned to me, to take him back to his parents' house and let him continue his studies at the technical college in Dresden. This was his deliverance; he became more sensible, passed his exams, and became a well-established manufacturer in Burgbrohl in the volcanic Eifel, where he exploits a natural carbonic acid spring, mainly for the preparation of alkali bicarbonate. He even was made a dignitary, for at his last visit to Berlin he related to me that he took part in an ecclesiastical conference as representative of the lower-Rhine synod.

We got on quite well in the provisional laboratory in Munich, despite manifold shortages of equipment. Our work met with success, which I will describe in detail in a forthcoming chapter, and Professor Baeyer did everything to encourage us and make our stay comfortable. Just as in Strassburg, I enjoyed his particular favor. Although I was a simple *Praktikant*, without any obligation to the institute, he dredged up various privileges for me that otherwise would be [available] only to *Assistenten*. We also became closer on a personal level. He often invited me as company into his family.

In the summer, when they went early to the country, we sometimes went together to a guesthouse. I learned much in conversations here outside the circle of scientific ideas. I clearly remember one conversation from the summer 1876, because I then made Baeyer aware of my intention to continue on a scientific career. To this end, I wanted to [return] to Strassburg the next winter in order to become better acquainted with the analytical instruction of Rose. He held this to be completely sensible, because Robert Bunsen was already too old, then laughingly added the remark that I am heading to my goal with great deliberation.

I used the autumn vacation, however, for a trip to Berlin, Copenhagen, and Hamburg. This trip was very enjoyable and turned out to be instructive for me as well. The first part of the journey led to Dresden and Leipzig. [I was] accompanied by cousin Otto. The Saxon capital greatly impressed us with the beautiful setting, the splendid buildings and the wonderful art collections. We also heard much about the residents from the previously-mentioned Andreae family, a couple with 10 children who lived [in Dresden] and very kindly received us cousins. Leipzig pleased us, not just as a university city, but also as a center of trade.

In Berlin, we stayed 8 days and met with Wilhelm Königs, who then joined us in our travels. We saw and experienced much more because of his large circle of friends. We were also quickly initiated into the secrets of night life in the big city. Nevertheless, the overall impression that the city left with me was not particularly friendly. As a consequence of the foundational period, the city was caught up in a transformational process, made apparent by huge development and many ugly buildings. But much also remained of the unsightly installations of old Berlin, and now I still remember with a shudder falling into a gutter that was at least 1/3 m deep and filled with a fluid that was by no means pleasant. Also, the fundamental feature of the population, their growling and impertinent manner of speaking, with the imperious tone reminiscent of the military, was quite strange to me. The chemical institutes that we saw aroused just as little of our acclamation. If someone at that time had prophesied that I would one day be appointed in Berlin, I would surely have energetically declined such a [hypothetical] offer.

From there, the journey took us to Stettin, where we immediately boarded the ship for Copenhagen. Cousin Ernst parted from us for this time and traveled in the second[-class] cabin, because he assumed the company would be better. It was my first sea voyage, and it has permanently remained in my memory, because it was quite stormy. Even in the lagoon many people were already seasick, and when we came to the open sea the small ship got into a severe rocking motion. Königs and I managed courageously until late evening, when the cold forced us to go into the cabin. Here we made our sacrifice to Neptune, and when we arrived at Copenhagen the next morning the entire ship's company was in a somewhat sorry condition.

The stay in the Danish capital, lying in a beautiful setting and dedicated to the art of Thorwaldsen, richly compensated us for the hardship we suffered.

The Danes are a very polite people, and in spite of the political aversion that they harbored against Germany at the time, they treated us with great civility. Only once did we collide with their prejudice. In a café, where we were the first guests, Königs pulled out a deck of cards that he always carried with him when traveling and wanted to start an innocent game of *Skat*. But the host raised resolute opposition, because in the better guest houses of the city cardplaying arouse offense.

Naturally, we also visited the beautiful surroundings of Copenhagen, the sea spa Klampenborg, the small city Helsingoer with the old castle well-known from Hamlet, and finally the splendid castle Frederiksborg. I never again came to Denmark; however, I have retained a permanently friendly memory of the beautiful maritime country.

The crossing to Kiel via Korsoer passed without incident, as the sea was calm. At that time, Kiel gave the impression of a small, dirty, somewhat unimportant sea town. Very little remained of the great German navy with the massive shipyard installations of modern times. Against that, on our arrival, which occurred early in the morning, we did have the pleasure of admiring the splendid Kiel bay. On the train trip from Kiel to Hamburg, we had just as welcome an opportunity to admire the pretty and pleasant land of Holstein, with its many small lakes and splendid forests.

Hamburg at that time was not yet the enormous commercial city that it is today. However, as a port it already assumed first-rank in Germany, and life at the harbor sprang a few surprises on us land rats. Also, the city itself, with the beautiful buildings on the Alster [River,] the splendid art museum and the good theaters was likely to attract the attention of travelers. As young people, who could spend the entire day sightseeing, we saw it all in a few days.

In the meantime, the conference of the German scientific researchers and medical doctors had begun, in which we, as members, took part. The chemical section was not very well attended; nevertheless, I did make several interesting acquaintances. First, A. Ladenburg, at that time Professor of Chemistry at the University of Kiel, then Associate Professor Michaelis of Karlsruhe, who just came from England and distinguished himself equally with a monster beard and a dilapidated top hat. Finally, Dr. R. Noelting, a friend and later brother-in-law of Witt. He was particularly interested in dyes and immediately asked me about rosaniline, about which I had made a small publication with the cousin Otto. At one session of the chemical division, I held my first scientific talk. It was somewhat short and dealt with asymmetric diphenylhydrazine, that I had recently found, and made interesting comparisons of its isomerism with [that of] hydrazobenzene. The president of the session, Professor Ladenburg, dedicated a few friendly words of recognition to me. Of the general sessions, that we likewise attended, only a presentation of the zoologist Moebius, from Kiel, remains in my memory, because this brought out a chemical curiosity. It was the identity of the mysterious protoplasm (Bathybius Haeckelii) with amorphous gypsum, which precipitates on mixing sea water with alcohol.

The mood of the natural researchers, "they research nature" in the local vernacular, was the best imaginable, elevated by the splendid autumn weather, the many opportunities for recreation in the maritime city and the excellent hospitality of the citizens of Hamburg. One celebration that remains in my memory is a steamer excursion to Blankenese. On the return trip we had the opportunity to witness the exuberance and rudeness of the harbor population. It was already half dark, and bunches of half-grown boys were entertaining themselves by tossing small but sharply exploding fireworks at the feet of our party, particularly the women. Somewhere else one might have become angry at such mischief, but the locals were used to it and remained calm.

From Hamburg we three traveled to the Rhine without stopping. In October, I parted company with the other two for one semester, as I went to Strassburg in order to [study] analytical chemistry with my former teacher F. Rose, and in particular, to gain experience in instruction of beginners. Naturally, I had already applied during the summer and stated my wish to be allowed to work with Rose as a volunteer assistant in the analytical division. However, the director of the institute, Professor Fittig, [gave me to understand] that he could grant my wish only if I would accept the post of a paid assistant. This I did, but a few months later had to experience the surprise of [learning] that my already quite modest salary would be reduced. It was a measure that perhaps was justified for administrative reasons, but nevertheless, it had a curious effect on the other employees of the institute, particularly the servants, because they viewed it as a degradation for me. Naturally, I just laughed over this, but it did fit with the personality of Fittig, who slightly annoyed other people through his directives, without meaning to do so. I have great respect for Fittig because of his excellent work, but in closer discourse with him I have noticed the great difference in comparison with Baeyer. In spite of his keen observational gift and his skill in experimentation as well as his [constructive] criticism, he dispensed with geniality and was not readily open to new ideas, for example, stereochemistry and the physical [chemistry] discipline. Although a good teacher for beginners, he did not have the capability to tie older chemists to himself and thereby build a scientifically greater school. In Tübingen he was accustomed to give the instruction in chemical analysis. He would have liked best to continue this in Strassburg, [but] the organization gathered by Baeyer was already in place. Among these was Associate Professor Rose, who was made independent leader of the [analytical] division. As Assistant, I was active only in this division.

Unfortunately, there remained for me a certain disappointment. When I was a student, I felt so much obliged to Rose for his instruction. Now it seemed one-sided to me, even with the many practical experiences of the Bunsen laboratory training method. New methods that appeared in the literature were not attempted at all, and the smallest deviation from the model [was] counted as an error. Nevertheless, it seems to me that what I learned from Rose's instruction later placed me in good stead when I myself had to take over the analytical division of the Munich laboratory.

The number of students at the Strassburg laboratory was not so large that my time was taken up entirely by instruction. The possibility remained for me, therefore, to do other things as well. I took part fairly regularly in the physics colloquium [led by] Kundt, and I made some studies of the morphology and physiology of the lower fungi. As already previously mentioned, I had the idea from Dr. A. Fitz, whom I had known in passing, and whom I now got to know better. He was a wealthy winemaker from the Palatinate, already mature in years, [and] unmarried. At that time, he was carrying out interesting experiments on bacterial fermentation, for which he made his name in the history of fermentation chemistry. Through him I learned of some publications of Pasteur, above all the book "Études sur la bière." (The practical consequence that it had for a brewery in Dortmund was previously described.) The chemistry of fungi interested me so much at the time, that if my stay at Strassburg were longer I would certainly have pursued my own research in this area. As it was, it was next necessary for me to acquire knowledge of morphology and to get practice with operation of the microscope and fungus cultures. This opportunity was made available by the botanical institute under the direction of de Bary, who was one of the most knowledgeable in Germany of the lower fungi. With little difficulty, I learned to recognize a series of molds and yeasts, mostly on dry culture media such as potato, turnip, [and] carrot. It has always remained a puzzle to me that only years later [did] Robert Koch introduce solid culture media for bacteria, and since then he is regarded as the inventor of these methods. It is a fact that Pasteur cultured microbes exclusively in liquid, that is, placed in his well-known flask. Also, his successors such as Fitz et al., never had the idea to use solid media.

Outside the laboratory I was not lacking for entertainment and company in Strassburg. As young people we allowed ourselves small pranks here and there. I remember one of these jokes. Dr. C. Wurster, who was serving out his one-year term of voluntary [military] service, was out at night barhopping and had overstayed his leave. To protect him from arrest by a patrol, we quickly transformed him on the street into a civilian.

[At Strassburg] I made a somewhat large number of new acquaintances, of which I will mention only two. First, the head of the apothecary at the city hospital, Dr. Musculus, a friend of Mering. He was considerably older than we, a native of Alsace, charming in company, and a good chemist. We have him to thank for the first synthesis of dextrin from grape sugar [glucose], with which I thoroughly busied myself 15 years later. The physicist Dr. Fuchs was a much odder character. He was a bar genius and a man of comprehensive education. He originally studied philology, then became a medical doctor. After completing all the necessary exams, he practiced briefly as a neurologist, then turned to physiology. This science also did not seem precise enough, so he became a physicist. Then, as a result of frequent conversations with me, he had the desire also to become a chemist. This did not occur, however, because in the meantime the modest inheritance from his father ran out, and he found it necessary to earn money. He later qualified in Bonn and, as I heard to my great pleasure, finally came to Essen as a personal physician to Friedrich Krupp. I can [easily] imagine that it brought pleasure to this creator of the greatest German industrial concern, at the end of a life rich in work, to converse with the multi-talented, witty, and childlike scholar.

In the spring 1877 I had achieved my goal in Strassburg and hastened to return to Munich in order to resume the work on hydrazine compounds and rosaniline, which [I had] left hanging. In the meantime [work on] the institute's new building was in progress, and [preparations] were underway to move out of the old building in the autumn of the same year.

The summer passed for us as its predecessors [had,] as an appropriate combination of diligent work and cheerful living. The number of students and older chemists had grown, and our social circle had also increased accordingly. We had all become good [citizens of] Munich and had adapted to the beer life. Particularly entertaining were evenings at the so-called beer cellars, that is, the bars of the large breweries in pretty gardens, some on the Theresian Way, some on the right bank of the Isar.

It was always merry there, a crowd drawn from all parts of the population enjoying the tasty beer. One brought along a bite to eat, the necessary purchases made at a butcher- and baker-shop along the way. One could see entire families assembled here, from small children to the aged. The beer life also had many other peculiarities. In certain areas of the city there were beer beggars, that is, individual people with an empty tankard who maintained that they had no money to buy beer, the aim being to have charitable people fill the tankard with small donations. The oddest outgrowth of the beer congeniality, however, developed at the bar of the Salvator beer, which is made at a single brewery in Munich and is typically drunk up at the Salvator bar after 5–6 weeks. I made my first visit in the company of Professor Baeyer. It was a frightful affair. In spite of the cold time of year, dozens of drunken people were lying around the garden and on the slope of the hill. Entry into the enormous closed space was fraught with difficulties, because troublemakers and drunks were continuously being thrown out. When we were finally inside, we were invited into a reserved room by a person whom Baeyer knew. A strange company was assembled there that one might have taken for lower-class people. The conversation was also uninformative. I still remember that one of the men pulled an old, strongly-smelling cheese from his pocket, cut it apart with the daggerlike knife that Bavarians always carry, and then invited his neighbors to consume it. On my inquiry as to the identity of this person, I learned that he was a high-ranking officer. My neighbor, whom I had also underestimated from his appearance, was the well-known sensitive poet Lingg. And so it gradually emerged, that the entire company were highly-esteemed men, artists, scholars, officers, high-ranking government officials, business moguls, and so on. At that time, I came to the conclusion that there exists in the world no means quite like beer that blurs rank and social standing and makes people equal.

A great advantage of Munich is the proximity of the mountains. At that time already one could take the train to Tegernsee or Lake Walchen in a few hours. We made use of the quite-frequent Catholic holidays of summer for many little tours of the mountains. At the Whitsun vacation, which lasted longer, we usually made our way to Tyrol, because the cuisine and wine were markedly better there. Neither I nor any of the other members of our circle were great mountaineers, but we often dared to scale the middle peaks, such as Herzogenstand [and] Wendelstein. Even here, under unfavorable conditions, danger can arise, which was made clear to me on the ascent of the Kitzbüchler Horn during a Whitsun outing. The peak of mountain was still covered with rather much snow. Our company consisted of perhaps 5 persons. Luckily, we had a porter, who also acted as a guide. When we arrived [at the top,] Dr. Boesler, about whom more later, took ill with mountain sickness so severe that he could no longer stand. We wrapped him in our coats and had to lay him in the snow. Then, to keep warm in the icy-cold wind, we [engaged in] vigorous exercise, calisthenics, etc. As the concern arose that the patient might no longer be able to start the return journey, we posed the question to the guide what we were to do. He replied calmly that he would stuff the gentleman in his rucksack and climb down the mountain. Luckily, Dr. Boesler was saved from this surely uncomfortable last resort, because he was soon recovered enough that he could make the return journey with our support.

On similar outings over the course of the 7-year Munich period, I [visited] nearly all of the beautiful peaks of Upper Bavaria and South Tyrol. Many of these remain in my memory, particularly a mountain trip that I undertook with the physiologist Dr. Tappeiner, the zoologist Dr. Graf, and a professor of anatomy at the veterinary school in Munich, Dr. Frank. [We] first took the train via Lake Starnberg to Pensburg, [the site of] the only coal mine in Bavaria, and from there on foot to Kochel. Here we ran into the autumn maneuvers of Bavarian troops, which had completely depleted the food supply. There was no bread left, but there was still plenty of beer. We had to stay in a hayloft in the company of soldiers. In the middle of the night, I fell softly into the cow stall underneath. The previous evening in the guest house a characteristic scene from Bavarian folklore played out before our eyes. Near us sat a man, probably a lumberjack, who, as later became known, made an attempt to cheat the barmaid. Suddenly, without a word, an empty beer mug flew from the bar at the head of this man. At the same time, a strong man stepped from the same corner—it was the so-called bar boy—beat the man with violent blows and disappeared quietly into the dark corner of the bar. The beaten man left the guest room just as quietly, and the matter was over.

The next morning, after a perfunctory wash at the spring, we climbed the Kochel-berg, then to Lake Walchen and from there, still on foot, to Partenkirchen. Professor Frank served as guide for our small company, for he had long served as veterinarian in these parts prior to his academic career and was therefore quite familiar with the land and people. We came in close contact with the population, and I had quite a good impression at that time from this fresh, energetic, and very naturally and freely behaving race. Unfortunately, the cooking, specifically in the guest houses of second and third rank that Frank led us to, left a great deal to be desired. Consequently, already on the fourth day in Partenkirchen, my stomach and bowel became thoroughly upset, and I had to return to Munich alone.

Another journey, which went better for me, I took with Dr. Fluegge, currently Professor of Hygiene at the University of Berlin. He had previously been a practicing physician at a health resort, then turned to the newly blossoming [field of] hygiene and came to Munich to work at the Pettenkofer school. We took a long walking tour, from Tegernsee to the Inn valley via Lake Achen and from there to Benedictbeuren via Mittenwald, Partenkirchen, Lake Walchen, and the Herzogenstand. He was a cheerful fellow, a brisk walker and [he kept his] eyes open for beautiful scenery and characteristics of the population. The conversation did not lack for scientific topics, and I attempted to convince Fluegge of the importance of Pasteur's investigations and of the meaning of the metabolic products of pathogenic microbes in infectious diseases.

In September of the same year the conference of the German scientific researchers and medical doctors took place in Munich. Baeyer had taken over the organization of the chemical division. Consequently, I was named secretary, together with Dr. Wilhelm von Miller of the technical college. The leadership ran into a few difficulties, however, because Mrs. Baeyer, who was on vacation in the Bavarian mountains, had just presented her husband with a little son, the current Professor of Physics at Berlin. Consequently, the husband appeared at the meeting with a small delay. The session was quite interesting for us chemists, because unusually many colleagues had streamed together. I first met: Victor Meyer, C. Liebermann, J. Wislicenus, F. Tiemann, C. Scheibler, C. A. Martius, and Peter Griess. The last [of these] was indisputably the oddest personality in our circle, as one will easily believe according to the most delightful necrologue, which A. W. Hofmann later dedicated to him. Due to overcrowding he had run into many inconveniences in Munich, particularly at the festive events. He peered in silently at the confusion for a time and later expressed his feelings in proper Hessian dialect: "The people don't understand crowd control."[1]

In the chemical section he was treated with great attentiveness and also chosen as chairman. The University of Munich afterwards bestowed the [honorary] title of Dr. phil. h. c., for just then it was the heyday of benzene chemistry, and the reactions discovered by Griess were being employed extensively to answer questions of interest. The azo-dye industry had taken a large upsurge. Finally, people who, like Griess, were in a practical profession and nevertheless used their free time for important scientific research were quite rare in Germany.

None of the talks of the chemical section has remained in my memory. Overall, the publication of scientific things does not play a large role at such meetings. Personal interaction of the participants is far more meaningful, as is the exchange of experiences that one would not confide in a public speech or a paper. It also gives the younger colleagues a welcome opportunity to present themselves to the older [colleagues] and demonstrate their skill in a talk. In later years I frequently attended meetings of scientific researchers primarily for the purpose of meeting younger colleagues. Many a friendship is [started] with just such casual conversation. For example, my friendly relations with Victor Meyer and F. Tiemann were initiated at this Munich meeting.

In the first general session, a talk by Haeckel caused quite a sensation, because he drew from advances in biology far-reaching conclusions for the entire intellectual and ethical life of the world, with sharp attacks on the church and state institutions of instruction. A few days later, in the second general session, Virchow answered him and led back to Haeckel's theories and challenges in interesting and learned ways. One had the general impression that only a few German scholars could compose, in the brief interval of a few days, so exemplary a scientific critique in the form of a popular speech.

At such a large gathering in Munich there was naturally no lack of festivities. Among others, the magistrate of the city had organized an informal welcome at the old town hall, where the political domain was also touched upon, which otherwise is just not customary at these sessions. An old Bavarian professor cordially welcomed the fairly large number of German-Swiss [who had] appeared, but then added a call for them to turn away from the French-Swiss and join the new German empire. The Swiss politely but quite firmly declined, with the appropriate remark, that the Swiss, according to their political and cultural structure as well as the historical development, see it as their duty to remain on friendly terms with all neighboring peoples and states.

---

[1] Translator's note: Fischer uses the word *Massen* here perhaps as a double *entendre*, possibly also intended to convey the scientific meaning.

In the chemical section my cousin and I were privately asked several times about the state of our work on rosaniline. We had to confess, with a certain shame, that we had made no progress for a year. This may well have spurred us on, for in the following winter semester [we did] in fact succeed [in finding] the solution. I also brought my investigation of hydrazine compounds to a certain close, in order that the collected discourse could appear in Liebig's *Annalen*. The immediate cause was my upcoming habilitation as *Privatdozent*. I was urged toward this [goal] by Professor Baeyer, because in the meantime the new institute had been completed and was ready to move in, and therefore the need for [lecturers] was becoming clear. Another colleague, Dr. Aronstein, had made known his intention to become *Privatdozent* at Munich. Since he was a man of only middling talents, Professor Baeyer [felt it important] that he not be seen as the new institute's first *Privatdozent*. To avoid this, I was shoved forward, although I would have much preferred to remain for a while in my independent position as private scholar.

My habilitation took place at the end of the winter semester in the spring 1878. Since the University of Munich did not recognize doctoral titles earned elsewhere as fully valid, I had to take yet another test in the form of a colloquium for the so-called naturalization of the doctoral title. As examiners, Professor Baeyer was appointed, as well as a second professor of chemistry from Liebig's time, Dr. Vogel, a very insignificant man. For Baeyer, who had known me for years, the test was merely a matter of etiquette, so he posed the question, seemingly in all seriousness: "*Herr Doktor*, can you let me know something about the compounds of hydrazine?" The answer was not difficult for the discoverer of hydrazine. And then, with the same gravity, he turned to ask Dr. Vogel, who apparently had never heard of hydrazine and also had not read my habilitation paper, "Are you satisfied?" That was the case. Vogel added to this avowal several somewhat foolish questions, whereupon the test concluded.

Much more difficult were the requirements of the habilitation itself. The faculty pose a theme which, after a pause of 3 days, must be discussed in a speech of perhaps ¾ hour. My theme read: "The current mission of chemistry." To this point I had never given a longer public speech, so one can well imagine that I had to work strenuously during the 3-day interval in order to achieve a workable speech. My memory stood me in good stead; for I was later able to [recite] the speech in the auditorium nearly word-for-word without the manuscript. Aside from the faculty, naturally acquaintances from the institute were gathered there, and many of them wished me good luck, that I might complete the speech without faltering. There followed a discussion of the thesis assigned to me, in which only members of the faculty participated. I had a brief encounter with the senior faculty member, the mineralogist Kobell, but after a small concession on my part, an accord was reached.

At this time, I don't want to omit to add that the faculty in Munich was sensibly divided into two sections. All business relevant to the natural researchers was considered only by the mathematics and natural research section. I hold this to be a great advantage over the Prussian universities, where the faculty are not divided and therefore must conduct all business *in toto*. For example, in Berlin the number of full professors has long exceeded half-a-hundred. This would make business meetings troublesome and require countless, lengthy meetings. It would also lead from time to time to unnecessary arguments over questions of principle between the representatives of the natural sciences and humanities.

So, I had then successfully become *Privatdozent* of chemistry and in fact the first in the new era of the University of Munich. In the summer semester the responsibility came to me to give a lecture [course]. I chose coal tar dyes as theme and had the good fortune to get a fairly large number of listeners, because such specialized lectures were rather rare at that time in Munich. The talk itself caused me difficulties from the outset. Although I was not used to speaking [without notes,] I [decided] on principle not to use a manuscript. However, this was possible only if I completely worked out the talk in advance and then memorized it. In this manner I was successful, after one semester, in [giving] such a talk [without notes] so that later it was necessary for me to prepare only the content and the association of ideas, whereas the manner of the lecture could be left [as extemporaneous.] Since it was often said of me, that my lecture was clear and the requirements of the listeners appropriate, so I believe I can recommend the process I adopted to young docents. It goes without saying that one must master the material, but the intellectual outline must also stand clearly before the eyes of the speaker. In my mature years it has pleased me to observe the facial expression of the listener, in order to determine whether the subject matter interests [him] and whether the tone was correctly chosen. Furthermore, it seemed to me sometimes permissible in the enthusiasm of the talk to undertake free excursions into neighboring areas or to develop ideas concerning unanswered questions that I myself have had and that I could pose only as hypotheses. When the listener notices that the speaker lends his name not only to things that are generally well-known, but also to a thing or two of his own intellectual production, then the personal feeling with the speaker becomes nearer, the attention closer, and the intellectual returns greater.

In the first lecture [course] at Munich I was certainly not blessed with the success of being in touch with the audience. After a few [lectures] a delegation appeared before me who declared in benevolent tones that they had understood neither the sense nor the words because of the all-too strong dialect of the speaker. I laughed heartily over this criticism and attempted to improve myself, and the audience rewarded me for it by appearing in comparatively greater numbers until the end of the semester.

I chose a much more difficult topic for my second lecture [course,] namely selected chapters of theoretical chemistry. At that time, the well-known book "Modern Theories of Chemistry" by Lothar Meyer had just appeared in a new edition, and as a fortunate complement, his brother the physicist had published a somewhat popular depiction of the kinetic theory of gases. I studied these books diligently and tried to present the content to my audience in a more concise and more popular form. This seemed to succeed, but in the continuation, I made a big mistake. I incorporated a critique in the remainder of the lecture material to separate the generally known from all hypothetical embellishment. The entirety amounted to a critique of the atomic theory according to the understanding at that time. For myself, this attempt was surely very instructive, but I caused disaster in the audience, and several of them could not hold back bitter complaints. "What are we to make of this," they said to me, "when we struggle to understand a professor's exposition of theoretical ideas, only to learn later that much of it is still doubtful." From this I learned that in lectures for students one must be very careful with critique, and one best present only things that are certain or apparently certain in the science.

In the spring 1879 Professor Volhard, the leader of the analytical division, was appointed [full] professor at Erlangen. At Baeyer's suggestion, the culture minister in Munich offered me an associate professorship at the university. If I were willing, I would assume the duties of Volhard at the chemical institute. I gladly declared myself ready and was installed associate professor in the faculty on April 1st the same year, with a salary of 3160 marks.

The news of the permanent position and of the first income of the dear son aroused great jubilation in Euskirchen. My father immediately wrote a congratulatory letter with the remark that he had celebrated the outcome, together with the mother, with a fine bottle of wine. He also inquired earnestly whether the new professorship could count on a large honorarium for lectures, as the salary did not seem to him very high. At the Bavarian ministry the opinion on this matter was certainly different, for the Head Director of the ministry Dr. Voelk had said to me that for such a sparkling salary, extraordinary efforts in teaching would be expected of me. I laughingly promised all of this but could not suppress the remark that there were opportunities for chemists to earn far more money than in an academic career.

With the assumption of the new position, that was primarily due to Baeyer's benevolence, my activities at the institute underwent a radical change. Until then I had been a free researcher, without any obligation to the university. In that capacity, it was possible for me to carry out all the necessary experiments for my scientific work alone. Only with the rosaniline research did I work with cousin Otto, and that played out very smoothly. From now on I could no longer manage in this way since the majority of my time would be devoted to practical instruction in chemical analysis. The help of assistants for private research would be indispensable. To be sure, I had already made a weak attempt at this one semester earlier, when I invited Dr. Erhardt to work with me on the mixed azo-compounds, particularly the phenylazoethane. But this help brought me little joy, since it was absent in all decisive things, and in the end, I had to do almost everything myself. It was therefore with a certain fear that I approached the choice of new co-workers. I also had the misfortune of making a mistake with the choice of the first private assistant. I allowed myself to be swayed by recommendation into accepting an outside man whom I did not know, Dr. Troschke from Berlin. After perhaps one year I was happy to let him go again, for he had more nearly hindered me than helped. After that my luck improved. Nearly without exception my private assistants have been splendid, diligent, and knowledgeable men, whose help I must recognize with the warmest thanks. The line begins with Magnus Boesler from Königsberg, a superb individual, who helped me first of all with the caffeine work and at the same time carried out his doctoral work on anisoin and cuminoin under my direction. Soon thereafter came Emil Besthorn from Frankfurt-on-the-Main, a man given to humor, whom Königs therefore singled out for his close circle at the bar.

The number of teaching assistants, who had their place in the large work halls, was much larger. At the apex was Clemens Zimmermann, a highly talented and extraordinarily [hard-working] chemist, who unfortunately died early. He was also gifted as a teacher. He completed his studies under Volhard and began his own investigations when I entered the analytical division. At that time, the periodic system of the elements was coming to be recognized more and more in inorganic chemistry, so it was natural that Zimmermann turned to tasks that were concerned with [the periodic system, for example,] the determination of the density of uranium tetrachloride and the specific heat of metallic uranium. If [his] health had been better, he would surely have enriched the chemistry of minerals with many handsome discoveries.

Two other teaching assistants who prepared their doctoral work under me were the gentlemen Lehnert and Renouf. The first later obtained an influential position as a member of the patent office in Berlin. The second became professor at Johns Hopkins University in Baltimore. He was quite a character in many ways and was distinguished by his forgetfulness. I often had to tell him which preparations were in his various bottles and dishes. He also loved the mountains and once had an accident when he was climbing alone, fell, and broke a leg. He lay there for 2 days without help, and the consequence was, that the subsequent healing left the bone much shorter. This did not dampen his enthusiasm for mountain climbing at all. He became a member of an alpine club, put on shorts and avidly hobbled up the mountains.

Lastly Krüss, an offshoot of the well-known family of opticians of Hamburg, was also active in the division. Like Zimmermann, he also perished early, indeed from pernicious anemia, which is extraordinarily rare among young men in Germany. I therefore suspect that it had more to do with a chronic poisoning, probably from hydrogen sulfide, for the ventilation in the institute left something to be desired; in the analytical division there was often quite a bad atmosphere for the majority of students.

As the University of Munich had only a single chemical institute, it was responsible for [training] not just chemists, but also many more apothecaries and medical doctors. During my time, the number of *Praktikanten* in the division, who mostly worked only half the day, was perhaps 150. Supervision of all of them was entrusted to me, so I could devote only a few minutes to each student, and even with this limitation it took me two days to make the rounds of both halls. My primary concern always had to be to urge the assistants to sensible activities. Also, I had students assemble in a large circle to hold a small lecture or improvise a test. In particular, the medical doctors were most grateful for this. Zimmermann adopted this system and developed it further with great skill.

Only a small number of the pupils at that time have remained permanently in my memory. Among these were Ludwig Knorr and Reisenegger, about whom more later, then a Mr. Ehrensberger, whom I assigned several experiments on the determination of arsenic in foodstuffs, following methods I developed, and on the determination of nitric acid as nitric oxide. The young man made a particularly good impression on me with his understanding, vigor, and lively interest for scientific things. When he made known his intention to become a teacher at *Gymnasium*, I laughingly replied that it would be a shame, he should remain in chemistry, and if he were obliged to earn money, he should go into industry.

He followed this advice and, through the intervention of an uncle, came into the service of the firm Friedrich Krupp. Here he made a laudable career, for he was later a very influential member of the directorate and fundamentally contributed to bringing the manufacture of steel at Krupp to the highest productivity. When I visited him in Essen 30 years after [his] studies in Munich, he was able, with sincere pride, to show me a part of the factory and above all the impressive scientific laboratory for chemical and physical investigations. I had already encountered him for the preparations of the building of an imperial chemical institute, where he always emphasized not only his personal [interest] but also the interest of his firm in the advancement of scientific chemistry. He has now retired from industrial work and wants to use the free time of his old age for astronomical observations at his property in Traunstein (upper Bavaria.) As a student he came back one day from a short trip to his hometown Berchtesgaden severely scuffed up. I jestingly inquired if he had taken part in a brawl, and he replied that he had suffered a fall, that was not completely without danger, while gathering flowers with a young lady. When I renewed our acquaintance I inquired about the fate of the young lady, and he laughingly replied that she became his wife and presented him with a large number of splendid children.

Another candidate for the teaching profession, G. Brandl, requested a topic for a small project that he might submit at the teaching exam. At that time, G. vom [*sic*] Rath, Professor of Minerology at Bonn, had just asked me to analyze some fluorine[-containing] minerals that he had collected, so I let Brandl use Wöhler's old technique for the determination of fluorine through conversion to silicon [tetra-]fluoride and its complete absorption in weighed vessels. We could then carry out the desired analyses with good success. The results were published in a paper of the Bavarian Academy of Sciences. Later, this Mr. Brandl, without my knowledge, did me a valuable favor in return. He recommended me most highly to Dr. Daller, the leader of the ultramontane majority of the state assembly in Munich, and thereby strongly influenced the approval of a new building for the chemical institute in Würzburg.

Among the medical students that I instructed at that time, there was also one who caught my attention through his talent and interest for chemical work. He later demonstrated his understanding of chemical problems through a noteworthy investigation of mucin, and he is now the famous professor of internal medicine at the University of Munich, Friedrich Müller.

For my own investigations there was a very fine private laboratory with side rooms at my disposal. In size, it corresponded exactly to the rooms that Baeyer used. Here I used all the free time left to me after instruction, not infrequently also Sundays, when in the winter one would thoroughly freeze, since the central heating was [then] at a standstill. Against that, evenings were dedicated to recovery, for anyone busy with experimental work during the entire day is generally too tired in the evening to make studies of the literature.

The city and the cozy lifestyle closely tied to it also invited one imperiously to company and merry conversation. For a while we fell, even as a small circle, into the English mealtime and ate as a quartet, that is, Dr. Königs, Dr. Tappeiner, my cousin, and I, at the notoriously elegant restaurant Schleich. Aside from us, the company consisted of two Serbian princes and the president of the Bavarian *Reichsratskammer*. Because of this custom, we were seen as ostentatious. In reality, though, it was a measure of time saving, for we remained at the institute the entire day, mostly from 8 o'clock in the morning until 5½ o'clock, and I got past my hunger with some bread-and-butter or through increased use of tobacco. After a while, though, this division of time was too strenuous, and we returned to the usual German midday [meal.] Yes, and I continued this habit even in Berlin, for whoever begins strenuous experimental work, that mostly must be carried out standing, between 8 and 9 o'clock needs a break at 1 or 2 o'clock, and this is best tied to the main mealtime. Then one can work anew 4–5 h, and according to my experience, the best use of the day for chemists [is made] when one let a series of operations run during the midday time that last long and do not need close attention.

The change of [meal] time also brought a change in table company for me. In the society of *Privatdozenten*, which held cozy meetings from time to time, I met many representatives of other disciplines and so I came into a new company in the art house. Among these was the philosopher Jodl, who later was professor in Vienna and made a good name for himself in his field. The two historians Stieve and von Druffel, both Westfalians, were also well taught and academically meritorious men, whose outlooks and goals interested me in many ways. Also, Stieve had a wonderful talent to make humorous speeches (so-called beer talk) of any length. His sarcastic wit and his inclination to find fault with the old professors contributed to the amusement of our young circle. We were all unmarried, until Stieve one day became engaged with a [young lady] from Bonn and then invited us occasionally. Through the historians I also came into contact with the artist family Kaulbach, where it was very merry. Later, after I had married in Würzburg, the first one to appear on the scene was friend Stieve, in order to check up on things, as he expressed it. All three are now dead.

Occasionally, in order to get completely simple healthy nourishment, I again ate at a beer house, the so-called *Abentum*, where it was good and inexpensive, and where hundreds of young people, particularly many artists, frequented. At our table there was a diverse company assembled: artists, scholars, technicians, young salesmen, and indeed there was even a genuine German prince among [us.] Here I came in closer contact with two art scholars. One was the *Privatdozent* of archeology Dr. Julius. His enthusiasm for the [goal] to transform chemists into appreciators of art increased to such a level that he held a private lecture for our little circle at the *Glyptothek* [museum.] This even had practical consequences. One artist, who belonged to [our circle,] decided to restore a famous, but badly damaged Nike, and I myself thereby came indirectly in contact with the plasterer of the museum. He asked me for a medium to remove black spots, which arose spontaneously on his plaster figures. I found such a medium in the action of chlorine gas, as it destroyed the spots, which were due to mildew, without damaging the figures. The plasterer applied this procedure on a large scale. He had to build a small booth for the operations with chlorine gas. How long this procedure remained in use or whether it was later replaced with a better one, I can't say.

The second art scholar at the beer table was Dr. Dehio, a Balt from Reval [Tallinn.] He later became a famous man through his great and splendidly designed works on the history of architecture. He finally became professor at Strassburg but has lived several years in retirement. I was very happy to see him again at Baden-Baden in the autumn 1917, after a long interval, and to learn that his wife is Paul Friedländer's sister. The events of the war in the East had moved him so strongly that he was occupied with setting forth a public propaganda [campaign] for the independence of the Baltic states from Russia and for [their] connection to Germany.

From the multiple interactions with artists and art scholars I gained more and more understanding of educational art, and multiple trips to Italy were well suited to strengthen these influences. This became apparent with me as well as with other members of our chemical circle through assiduous visits of the art exhibits that were organized at the Glass Palace, which was right next to our institute. In the last years of the Munich stay I also went fairly regularly on Sunday afternoons to the exhibit hall of the Artists' Society under the arcades of the Court Garden.

One became quite aware what a favorable atmosphere prevailed in Munich for artists. The majority of the educated population were interested in their works, took part in their development and their personal fates, but avoided spoiling them with overblown extravagant praise or ostentatious invitations to dinner parties and the like. Also, the life that the artists led among each other offered many advantages. In general, one heard the teacher praise, even if the point of view of the youth with regard to the goals and the media of the art were not infrequently quite different. The realistic and open-air painting technique, often exaggerated to the point of caricature, which, if I am not mistaken, originated in France, just began to take hold in Munich. I still remember the sensation caused by a picture by the youthful Max Liebermann "Christ in the Temple" at an exhibit in Munich. For those people who adhered to the old system, it was [seen] as an aberration, and for the church-oriented organs it was described as a blasphemy. But a large number of the young artists and also we young chemists, who were not uninfluenced by them, saw an earnest endeavor to adapt painting to reality and not to seek the spiritual penetration of the object in an exaggerated idealization of the figure.

The public festivals organized by the entire community of artists in Munich were justly famous. I attended two of these. One was a camp from the time of the 30-Years' War that played out in the *Grosshesselohe* Forest and gave a delightful picture of the armed force 400 years ago.[2]

The second was a costume festival in Munich that unfortunately experienced a tragic ending, the result of an accident with fire. One of the largest halls in Munich was transformed into a type of fun-fair through extraordinarily talented construction developments. One could enjoy everything that one expects to see at such an opportunity. Entry was permitted only to gentlemen, who naturally for the most part were disguised as women of every type. Consequently, the artistic imagination could be indulged without any boundary up to eccentric excess. I have seen such

---

[2] Translator's note: The Thirty Years' War, fought from 1618–1648, began 300 years before Fischer wrote this memoir, not 400 years.

crazy exuberance at Carnival more than once in Cologne, where [all of] the artists, [ranging in age] up to the highest age, seemingly were gathered completely. One delightful costume idea that was superbly realized has remained in my memory. The well-known painter Piglheim came as a prince, naturally with retinue. He paraded around before the committee, naturally entirely in the style customary for a princely visit. A short time afterward, there appeared an actual Bavarian prince who was received in exactly the same way. To the great enjoyment of the entire assembly the two trains met. The actual prince had the good enough [sense of] humor to greet the pseudo colleague in a friendly [manner] and they carried on together with the review.

Unfortunately, the festival was disturbed in the later hours by a frightful accident. Perhaps a dozen pupils of the arts academy came as a group of Eskimos. For this purpose, they had built a special hut and donned Eskimo suits that were made entirely of oakum. [In making these suits,] one had carelessly omitted to impregnate the material chemically to render it fireproof. Also, the hut itself was filled with inflammable materials of every kind. Incidentally, the majority of the structures had been constructed with similar carelessness. Since smoking was ubiquitous [there,] we chemists immediately had the sense of the highest danger of fire. I remember that we withdrew fairly early to a safe beer cellar and only from time to time looked out at the goings-on in the main hall. Suddenly, it was said, the Eskimos were on fire, and we in fact witnessed these poor persons running around on fire. They were fairly quickly extinguished outside their hut, and several of them believed, after coming through the danger, that they should compensate by having a generous drink of beer, but the injuries were so severe, that to my knowledge they all died in the next days. Luckily, the fire in the Eskimo hut was extinguished. Had it spread and the neighboring very inflammable objects caught [fire,] hundreds of people would have died, for the accesses to the hall had been narrowed so much by the construction that it would have been impossible to save the majority of them.

At all such festivals the supervisory body should always demand most rigorously that all necessary measures be taken against any danger of fire. The public is inclined to regard such directives of the authorities as arbitrary and punctilious annoyances, but anyone who has experienced a fire accident such as occurred at the costume festival in Munich and whoever, [such] as we chemists, knows how rapidly fire can develop will in my view gladly consent.

Aside from the formative art, music and theater also played a large role in Munich, and I must confess that I had great enjoyment from both.

Unfortunately, I had already neglected piano playing in Strassburg, and in Munich I did not risk beginning again, because a professor of music, who was a virtuoso, lived in our house. There was also the sense that in large rooming houses, one can become a real annoyance to many people through mediocre play. I preferred, therefore, to listen to good music from time to time by attending the excellent Odeon concerts or the Royal Opera. The latter was endowed at that time with outstanding strength. Wagner's music was particularly favored. Although the old classical works of Mozart, Beethoven, Maria von Weber, Lortzing, etc., were closer to my understanding, nevertheless the superb Munich production of the great operas of Wagner, for example, the "Ring of the Nibelungs," made a powerful impression on me. This grew yet stronger with a visit to Bayreuth, which I later undertook with several friends from Erlangen, and where we were present for a performance of "Parsival."

The married couple Vogel dazzled in the Wagner operas at that time in Munich. But also, no listener will have forgotten the old Kindermann with his durable powerful baritone, who mingled with the Munich public just as with old acquaintances.

I had even more enjoyment from good theater, which likewise was typical in Munich. The Munich theater had the good custom, from time to time, to present a series of works by the same master at reduced prices. One such series, which comprised the Shakespeare king dramas, was so powerful in its presentation for us young people, that a serious disturbance prevailed for several weeks over the experimental work in the laboratory. I was so moved by many scenes that bright tears flowed over my face, and I completely forgot my surroundings. This strong sentiment for dramatic charms has remained with me into old age, and my late wife said to me more than once that my lack of self-control at the theater was uncomfortable for her.

In attending theater, concerts, or the like, where people sit together densely, it can easily happen to a chemist that he gives offense through the bad smells that have settled on his clothes or hair. I was never blamed for this when we [took] the cheap seats, which we did as young people. But I once had to give up an orchestra seat at the Munich opera, because the neighbors were so bothered by the fumes streaming from me. The chemist should also be aware of this possibility in private company, and in case he has just [worked] with strong-smelling materials, he should take particular care with cleaning the body and the suit.

Unfortunately, sport came up the loser in the Munich lifestyle. We had no time for it and were also too tired in the evenings. A certain compensation came during the autumn vacation, which I always spent partly at Euskirchen and, together with my father, assiduously used for hunting.

As to social interaction as it played out in the family, I didn't do much, partly from convenience, partly because of a certain clumsiness in the company of ladies. By the way, the opportunity in Munich was ample, for Mrs. Baeyer led a large house, organized attractive parties from time to time, and took the greatest pains to draw in us young chemists. To a certain degree, she played [the role of] overall mother of the laboratory and was always ready to help when one of us was in need due to sickness or other cause. At her house I also met a large number of interesting people. Above all, she always invited us to table when chemists visited from [out of town.] After the fashion of helpful and capable ladies Mrs. Baeyer would also have gladly married us off, which she believed she could count on, since she knew how to gather a circle of pretty and charming young ladies around her. As far as I know, aside from [her] own daughter, she did not succeed in [marrying even] one of these misses to a chemical man. The young chemists gathered around Mrs. Baeyer at that time to a man were unmarried, and also strikingly many of them remained in this condition, [which was] not to the benefit either of themselves or the science.

My work on hydrazine and rosaniline had found outside recognition. This first showed in a call I received in spring 1880 from the Aachen polytechnic school. Landolt had been appointed to the agricultural college in Berlin, and in his place two [full] professors would be named for inorganic and organic chemistry.

The inorganic position with the large experimental lecture was intended for the assistant [professor] of the institute Professor Classen, and the professorship for organic was offered to me by the minister of culture in Berlin. In order to learn more about the position, I traveled to Aachen and conferred with the colleagues Professor Stahlschmidt and Professor Classen, as well as with the director of the institute Mr. von Kaven. Naturally, I also visited Landolt, who received me in his bed, because he suffered from severe gout. A splendid new building had been erected for the chemical institute, because the Aachen-Munich fire insurance company had made [large resources] available. But the operating funds were comparatively limited. No one would give me any assurance in this connection, and the entire atmosphere at the polytechnic school seemed to me so unattractive, that despite the keen wish of my mother to have me once again in the vicinity, I flatly declined the position.

Next, I experienced a most unusual [occurrence.] After perhaps six weeks the call from Berlin was renewed. The department head concerned at the culture ministry, Dr. Wehrenpfennig, wrote me to say that I surely did not properly understand the intention of the ministry. As I later learned, he had in the meantime spoken with his friend Dr. von Brüning from Frankfurt-on-the-Main about my appointment and complained that I had declined the handsome appointment so coolly. When the latter replied that I might have greater pretensions than what was offered me, [he] decided to make a second attempt to win me over. I was then received in Berlin by Wehrenpfennig and Landolt, who had in the meantime moved [to Berlin,] in the most engaging manner. [They] promised to comply with all of my wishes. But the aversion to Aachen remained, and as also Baeyer advised me not in the least to move to the polytechnic, I declined a second time, and I have also not once regretted the decision. In my place Professor Michaelis then came there.

Chapter 6
Erlangen

Two years later the professorship of chemistry at the University of Erlangen again became open, because Volhard went to Halle as successor to Heinz. I received the call and accepted. Indeed, the place held few attractions, but [the position] was full professor at a university, and the institute was comparatively well equipped with a new building. On a visit to Erlangen, I was also received by colleagues in a very friendly [manner,] and Volhard took pains to make clear to me the advantages of the new position.

Before they let me go at Munich, I naturally had to be offered several celebrations. The Chemical Society, of which I was the president, organized a modest farewell party, for which Königs composed the previously-mentioned Guano Song. But also the students did not want to miss the opportunity to hold a drinking bout. After the experiences on the occasion of my call to Aachen, where I nearly received blows from the drunken chemists, I was not very much inclined to expose myself yet again to this possibility. Only the threat that, by declining the invitation, I would give offense to the chemical youth, with which I stood on very firm ground, made me accept. So, then the celebration took place at a beer cellar, and it passed without incident for me, because I took my leave early. But the next morning I had to [hear the news] that several young people who wanted to go to Erlangen with me had been thrown out of the beer cellar.

At the beginning of April, I moved to the Franconian college. As it happened, in Nürnberg Miss Agnes Gerlach, my later wife, together with her father, climbed aboard the fast train and into the compartment where I was sitting, and we therefore arrived in Erlangen together.

[Finding] quarters in Erlangen was not particularly easy. First of all, a suitable apartment was lacking, and also the guest houses of the poor city offered no comfortable situation. Therefore, I first lay down in the spacious conference room of the chemical institute, where a very useful laboratory servant by the name of Griesinger saw to my needs. For food I was directed, however, to the guest house where a large circle of professors and *Privatdozenten* gathered at table. I had lived ten years in guest houses, so I felt the need to begin my own household. To this end I had already

© The Author(s), under exclusive license to Springer Nature Switzerland AG 2022
D. M. Behrman and E. J. Behrman, *Emil Fischer's "From My Life"*,
Springer Biographies, https://doi.org/10.1007/978-3-031-05156-2_6

ordered furniture in Munich. To protect me from being cheated, my sister Emma made a special trip from Rheydt to Munich with her daughter Hedwig. They stayed with Mrs. Baeyer, who immediately took them in as overnight guests.

The [delay] in finding an apartment in Erlangen did not last too long. After perhaps two months, all the furniture arrived and with it my first housekeeper, a widow from Westfalia, who had been sought by my sister Bertha and my mother. Thus, I moved into the first furnished apartment with an independent household in one of the few distinguished houses in Erlangen, which had earlier served the officer of the margrave's court. The apartment had a large hall with marble flooring and plastering in the rococo style. [The hall] later became famous in the debate over my work, for I performed the odor experiment with mercaptan here.

In the meantime, the work in the laboratory and the lectures had begun. Luckily, two splendid young men, Knorr and Reisenegger, who had not yet finished their doctoral degrees, had followed me from Munich, and I was able to assign them assistantships without a [second] thought. We quickly became friends. It was too boring for me to eat at home alone, so I invited these two young men to be my companions at table for at least the midday [meal.] This they gladly did, because my housekeeper was an outstanding cook, and also there was no lack of good wine.

Both had received their analytical training under my direction at Munich, and Knorr had also been occupied with the preparation of a dissertation on piperyl hydrazine when we moved to Erlangen. Reisenegger soon thereafter received a dissertation topic from me: "The reaction between phenylhydrazine and the ketones." Both successfully completed their doctoral degrees in Erlangen. Reisenegger remained for a time as assistant, then went to industry and, in particular, to Meister Lucius & Brüning in Höchst-on-the-Main. He was originally from Upper Bavaria, if I remember correctly, from Murnau, wanted originally to become an officer and completed cadet training in Munich. He changed to chemistry following a bodily injury. At Höchst he progressed through all levels as a technical chemist to become a member of the board of directors. A few years ago, in his maturity, he became professor of technology at the technical college in Charlottenburg, the successor to Witt.

Ludwig Knorr stayed with me longer, as he decided from the outset to make an academic career. He stemmed from a wealthy Munich family in business at the well-known firm Angelo Sabadini in the Kaufinger Street. He distinguished himself as a youth, not only through cleverness, experimental skill, and eloquence, but also through special personal amiability. In addition, he embraced all possible sports, was an excellent dancer, and played an important role during the Munich time in society there, for example, also at Baeyer's house. In addition, he succeeded, despite his youth and his still inchoate career, to win [the affections of] the pretty, much sought-after Miss Elisabeth Piloty, daughter of the great historical painter of Munich, who became his wife in spring 1884. The couple later moved to Würzburg with me. I have experienced cordial friendship with them up until today, so I will return to them later.

In the laboratory in Erlangen, we met the *Privatdozent* of chemistry, Dr. Vongerichten, a native of the Palatinate. He had worked for a time in Baeyer's laboratory and was therefore well-known to me. I was pleased to be able to entrust to him a place in the attractively furnished private laboratory. There he carried out his important experiments on the transformation of morphine into phenanthrene and thereby ushered in a new epoch in the chemistry of alkaloids. I also entered into quite friendly relations with him. But after perhaps three semesters he left the university and joined the chemical company Meister Lucius and Brüning in Höchst-on-the-Main, where he was employed in the scientific laboratory. His employment served the factory well, for it came into possession indirectly thereby of the patent for Antipyrine.

In the early time [in Erlangen] there were very few students in the institute. Volhard had not worked to attract students, and it was said that he was too strict on exams.

Holding the large experimental lectures at first gave me quite a bit of trouble, for the many experiments, which are indispensable, require precise preparation for the professor as well as for the assistants. Although the good book by K Heumann on lecture experiments was already in existence, I felt it necessary to create a special book with all the details for this purpose. This [book] moved with me to Würzburg and Berlin, and naturally over the course of time the contents were enriched. Also, in the laboratory the entire [responsibility] for instruction fell on my shoulders, because only young and inexperienced assistants were available. Then there were tests and the faculty business and also, of course, my own investigations.

Unfortunately, the technical furnishings of the institute were insufficient, above all the ventilation. My assistants and I frequently worked with phosphorus pentachloride and suffered severely due to [inadequate ventilation.] I will return to the question of ventilation in the detailed discussion of the new buildings in Würzburg and Berlin.

Over the course of several semesters the attendance at laboratory again became satisfactory. In the analytical division considerably many apothecaries, a few medical doctors and chemists had assembled, and the small organic division was already so oversubscribed that I had to reclaim a part of the old building that was no longer being used.

Of the new chemists, three earn special mention. Dr. Ernst Täuber, from Silesia, helped me as a private assistant in the preparation of acetone bases. Kužel was a teaching assistant in the analytical division. I worked with him on the hydrazine [derivative] of cinnamic acid, indazole, and benzoyl acetone.[1]

Finally, Julius Tafel investigated the isometry of indazole and isoindazole and took part in my work on sugars at Würzburg.

Täuber later came to the technical institute in Berlin and is also a member of the patent office. Kužel, a man privileged physically as well as intellectually, had quite a successful career at the factory of Meister Lucius and Brüning but later returned to Vienna, his home city, and made a name for himself in electrical technology through the production of metal filaments from colloidal metal.

Julius Tafel remained true to the academic career. He became *Privatdozent* at Würzburg and later full professor and director of the institute, as successor to Hantzsch. Unfortunately, lung disease forced him early to give up his successful

---

[1] Translator's note: Indazole is formed by heating o-hydrazino cinnamic acid. [Fischer] The work on benzoyl acetone was unrelated.

scientific work, particularly the electrolytic reduction of organic compounds, and to live only [for his] health.

Of the other coworkers I mention O. Bülow; the lecture assistant Koch, who prepared trimethylene diamine and worked on the syntheses of urea from such diamines; Elbers, who made the first hydrazine acids; and Hess, who took part in the syntheses of indole derivatives from hydrazine. Then Hegel, a grandson of the philosopher, who is now a member of the patent office, Roese, and finally Antrick, currently director of the publicly traded chemical factory, formerly Schering, in Berlin.

C. Paal and O. Hinsberg occupy a certain special position because they worked on topics that did not originate with me, but [they] gladly accepted theoretical and experimental advice and help from me. The first carried out his interesting experiments on acetonyl acetone and its transformation to furan derivatives. He later became professor of pharmaceutical and applied chemistry in Erlangen and then also in Leipzig, as successor to E. Beckmann.

Hinsberg came from Göttingen and discovered quinoxaline at Erlangen. His publication, which was dated from the Erlanger institute, precipitated an attack on me from G. Körner in the treatises of the Roman Academy. Körner, whom I learned of much too late, maintained there that he had produced quinoxaline before Hinsberg and had made this known to me orally during a visit to Milan. I did not respond publicly to Körner's reproach, but I did write him privately and made it clear that his complaint was unjustified. I could not in any manner remember such a communication, but if it really did occur, then Körner would have had several years to publish his observations. To make a priority claim after so long an interval, based on a supposed private communication which was not substantiated in any way, and then to raise an accusation of indiscretion against a colleague is a very questionable form of polemic, which certainly no reasonable researcher would recognize [as legitimate.] I can only bear witness here to [the fact that] Hinsberg was not influenced in the slightest by the alleged observation of Körner.

When I add that in Erlangen I discovered the osazones of sugars and thereby laid the groundwork for my later work on sugar, further that I obtained the first oxypurine through treatment of methyl uric acid with phosphorus pentachloride, further that L Knorr discovered Antipyrin, where he used the reaction I had already casually described between phenylhydrazine and acetoacetic ether in a very ingenious manner, one will get the impression that we were diligently at work. Only the evenings remained for social life.

The city offered little, here and there a concert organized by music-loving men. We once had the pleasure at such an opportunity to hear Bülow in Erlangen with his superb orchestra. Then occasionally there was also a production in the old theater of the margrave that now belongs to the city and where a Nürnberg troupe customarily gives guest performances. For more of these pleasures, one had to drive to Nürnberg nearby.

The confinement of the small city led the members of the university to inner consolidation. Interaction among families was eagerly maintained. For myself, I profited extensively [in the company of] my friend Wilhelm Leube and his wife

Natalie, a daughter of the excellent chemist Adolf Strecker. Naturally, Mrs. Leube called on her friends when necessary to help with the organization of social events. For a ball that Leube as prorector of the university gave, I had to undertake the production of a giant bowl [of punch.] I used the opportunity, in order to produce a carnivalistic procession, to glorify the pouring of wine. The main character was the strong, glowing servant of the institute Griesinger, who appeared as the master wine steward and entered the ballroom with a huge barrel [loaded on] a decorated wagon. On the barrel, dressed as Bacchus, sat a handsome ten-year-old boy, the son of the surgeon Heinecke, and around the barrel sat Leube's three daughters, [dressed] as nymphs. At this point the entire company started to sing the well-known song of the Rhine wine: "adorned with leaves, the lovely full goblet..." The punch, which I made from 150 bottles of wine, brought applause and was drunk up in a few hours.

Most frequently, we bachelors naturally met at the guest house, where a very cozy tone prevailed.

Among the young medical doctors, the closest to me were Leube's two assistants Penzoldt and Fleischer. With the first I carried out several experiments, in particular the one concerning sensitivity of the sense of smell. He is now full professor of internal medicine at Erlangen. Fleischer's fate was less lucky.

Among the older medical doctors, Leube was perhaps the most prominent personality. His specialty as gastroenterologist led to countless patients in the Erlanger clinic, including many interesting people.

I am glad to remember the other medical doctors: the pathological anatomist Zenker, the physiologist Rosenthal, the surgeon Heinecke, the gynecologist Zweifel, and above all the senior [member] of the faculty, the anatomist J. von Gerlach, my later dear father-in-law. He played a large role in the medical faculty, not only because of his scientific renown and honors as a teacher, but also because of his engaging personality and his efforts for the well-being and regard of the faculty.

We chemists were quite close with the associate professor of pharmacology Filehne, who first determined the fever-reducing action of Antipyrin and secured its entry into practical medicine.

Among the lawyers, Professor Marquardsen became the best known to the public as a politician. For many years he was a member of the *Reichstag* as well as the Bavarian *Landtag*. Later, in Berlin, I repeatedly had the honor and pleasure to see the old experienced, clever, and cheerful gentleman in my house. When my wife was alive, he also brought along his family, who were close friends with the Gerlachs in Erlangen. The second jurist with whom I often came into contact was the Swabian Hoelder, who later came to Leipzig.

Theology was represented in Erlangen only by an evangelical faculty, and [they] stressed their Lutheran character with great determination. Indeed, a representative of the Calvinist community of the Palatinate was denied membership. In the church world, the faculty enjoyed a good reputation. This was evident from the large number of young theologians from northern Germany who completed at least a part of their studies here and at the same time found advantages in the very inexpensive but also very simple lifestyle of the south German small city. The custom of young theologians to become engaged early attracted girls wishing to marry, and so Erlangen had become the refuge of more than 100 pastor widows who hoped to marry off their daughters. To make fun of this, a loose association of old, somewhat lazy fraternity students, who believed passing exams in Erlangen would be easier, adopted the name "Pastors' Daughters."

In contrast to Würzburg and Munich, the faculty was not divided in Erlangen and was led in important matters without apparent antagonism by its senior [member], the historian Hegel, son of the philosopher. Among the natural researchers, I mention the physicist Lommel, a sensible and tranquilly inclined native of the Palatinate; the botanist Rees, who carried out an excellent study of yeast under de Bary but restricted himself in Erlangen to teaching and social concerns; then the talented and witty zoologist Selenka, who later undertook extended travels in Java with his wife; the very meritorious mathematician Noether; his outwardly comical special colleague Jordan; and finally the professor of pharmacy and applied chemistry Albert Hilger. He was naturally the closest to me, and we also arranged a chemical colloquium together.

Hilger had the ambition to tackle chemical problems and also had the goal of publishing a new edition of the well-known works of Husemann on plant materials. But I soon became convinced that his talent and training were not adequate for these things, and so I advised him to turn to practical things, namely, the analysis of foodstuffs. He took my advice and had the best success, for it is attributable to his efforts that in Bavaria investigations into foodstuffs was transferred to the institutes for applied chemistry at the three state universities and that the control of foodstuffs in Bavaria was arranged much earlier and better than in north Germany, particularly in Prussia. Hilger was musical and, with his clever wife, a native of Holland, provided for the maintenance of the musical life in Erlangen. He later came to the University of Munich, where he built a new building for pharmaceutical and applied chemistry in the vicinity of Baeyer's laboratory.

In Erlangen the students naturally played a key role, because a large part of the inhabitants depended on them for [their living.] Also, the fraternities thrived in unusual manner, and in that connection dueling, despite the legal prohibition, was seen by everyone as natural and necessary. Yes, I myself, together with Penzoldt, Knorr, and a few other young people, took part in a course on fencing with saber. Naturally, this was only for the joy of bodily exercise, but to our surprise, we later heard that for Knorr it was a preliminary exercise for a student duel that he had to fight out with Dr. Friedländer in Munich. As I also often visited the university's swimming pool, due to the pleasure I took from swimming, I was done the unexpected honor when I was named by the Senate as member of the Fencing and Swimming Commission. As such, I could take an active part in the selection of a new fencing master.

In the year 1883 a change of personnel took place in the direction of the aniline and soda factory of Baden. First of all, Dr. Heinrich Caro, the previous leader of the scientific laboratory, wanted to retire or join the board of directors of the factory. It was attributed to his influence, that the principal stockholder and chairman of the board of directors, Mr. Sigl from Stuttgart, made me the proposal to become Dr. Caro's successor. Although this position brought in much more materially than any professorship in Germany, the academic career, with complete freedom of scientific work, was more to my liking. I therefore declined the position but gladly accepted an invitation to make a two-week visit to the factory, partly out of interest in the industry of coal tar dyes, and partly in the hope of being able to obtain raw materials in large quantities for my investigations. And so, I went in August 1883 to Ludwigshafen-Mannheim.

The 14-day stay was a fortunate combination of zealous work and merry company. I was first led through the entire factory. In every section, I was instructed by the pertinent leader in the details of the operation in familiar but very extensive manner. This was preferential treatment, very seldom allotted to academic chemists, and I was grateful for it, [and] that it was the management's wish to remain in association with me. At the same time a considerable number of experiments began in the technical laboratory on the methylation of uric acid, following the specification that I had worked out in detail in Erlangen.

As raw material, I had obtained several kilos of snake excrement from Amsterdam through the intervention of Professor Forster. On my inquiry as to why the material was so expensive, I received the surprising answer that in Holland it is used for the production of secret medicinal remedies and therefore [commands] an appropriate market price. My previous private assistant, Dr. Boesler, assumed oversight of the methylation experiments, which required perhaps 10 days, so that I scarcely needed to attend [to them.] It was different with a second preparation, o-amino cinnamic acid, which I made from a material in plentiful supply at the factory, o-nitro cinnamic acid, by reduction with ferrous sulfate and ammonia, following Tiemann. The operation had to be carried out in very dilute solution and required filtration of a thick sludge of iron hydroxide. For this purpose, an appropriate operational apparatus of the azo-dye factory was placed at my disposal. During this work, which I had to undertake myself, the strict prohibition of smoking in the factory was distressing to me. But when the old Engelhorn, a member of the management, visited me, we each lit a cigar during conversation. We were soon caught by the operations director, Dr. Burkhardt, an old Munich acquaintance of mine. A great row ensued, and we were ruthlessly shown the door with our cigars.

I used afternoons either for small outings or for hunting woodcock in the fruitful and game-rich vicinity of Ludwigshafen, where I was taken in a friendly gesture by the board member Dr. Karl Klemm. Evenings I regularly spent in the large chemical circle at the *Pfälzerhof* in Mannheim, which was distinguished by the first-rate Palatinate wines [served] at the bar. The technical men were grateful that, through many amusing stories, I gave them information about advancements in the science in broader form than they could otherwise experience. The suggestion was jestingly made that I be affiliated permanently with the factory as "Lecturing Advisor." In fact, at the end of my visit, the management proposed to me a contractual arrangement with the factory, for a moderate annual salary, whereby I would have only the responsibility to concede the rights to sell any inventions made in the areas of coal tar dyes and drugs. This same offer was made to Victor Meyer and Adolf von Baeyer. However, it was never realized, because soon thereafter a large change in management personnel took place at the factory.

After completion of my work, I traveled from Mannheim to Euskirchen, in order to assist my father with the woodcock hunt. Here I had the pleasure of receiving a visit from Victor Meyer. Roughly ½ year earlier, at Baeyer's house in Munich, he had expressed the wish to get to know the hunt, for this sport had always remained foreign to him. At that time, I invited him to come to the lower Rhine during the autumn vacation. He now made good on his assent. Unfortunately, he had just been on a strenuous mountain climbing expedition in the region of Bernina, came directly from the fresh heights and was somewhat fatigued at the warmer [lower] elevation. Consequently, his bodily strength was not sufficient to take part with pleasure in the somewhat strenuous hunt for woodcock. While my father and I could stride 6–8 h through the potato fields without tiring, he was generally lying exhausted behind a bush after 2 h.

A small hunt that I organized for him in the Flamersheim forest was therefore easier and more interesting for him. He had the remarkable luck, over the course of perhaps 2 h, to run across a buck and 5 wild sows. To be sure, he did not succeed in bagging a single animal. At the preparation of the midday meal that our forest ranger attended to, all members of our small circle took part in peeling potatoes, subject to the condition that each [man] should peel as many potatoes as he intended to eat. At this curious custom, Meyer laughed emphatically and confessed that up until then he had never in his life peeled a potato.

The day was less strenuous, and this style of the hunt suited him better than scouring the open field. But apparently, he now held his experiences at the hunt to be sufficient, and so I suggested to him an outing to the volcanic Eifel. He agreed, and we now traveled partly by wagon, partly on foot via Rheinbach to the Ahrthal, from there downhill to the Rhine, and then through the scenically pretty and geologically very interesting Brohlthal to the lovely Laacher Lake, one of the most beautiful volcanic lakes in the Eifel. This naturally interested Meyer to a high degree. In order to show him one of the characteristic carbonic acid springs of the region, I led him the next day to Burgbrohl to [meet] my cousin Dr. Hans Andreae, whom I knew well from the Munich time, and who had landed here as manufacturer and father after a stormy time as a student. He mostly produced pure alkali bicarbonate [from] natural carbonic acid that came out of the earth in mighty streams of gas as so-called mofettes and was collected for the factory using a metal pipe. The spring now serves for the production of liquid carbonic acid, which has since become a meaningful item for trade.

Naturally, we also paid a visit to the beautiful Jesuit cloister Marialaach on the lake, then returned to Euskirchen in the most magnificent weather. Meyer wanted to take leave of my father, with whom he had quickly made friends. In the meantime, an invitation from Wilhelm Königs had arrived. [He invited us] to visit Cologne, where he wanted to show Meyer the many beauties of the old trading city. [The invitation] was accepted.

On the drive Königs received us already in Kierberg near Brühl with the invitation to stop over at the country house of his brother, the Cologne banker. He had received the commission from his sister-in-law, an intellectually very lively woman, because she wanted to meet the two professors. At a merry breakfast, with the liveliness of a

Rhineland woman, she knew how to interrogate us in detail on our scientific goals. She rewarded us with sumptuous hospitality and by showing us her own works of art, a series of tolerably painted oil pictures. At this opportunity I saw my later pupil Ernst Königs, who is currently *Privatdozent* in Breslau, as a boy for the first time. Meyer and I then parted [company,] and he drove via Cologne to Berlin to visit his parents.

Before that, in Euskirchen, we had, at his suggestion, agreed to a close friendship, and I must confess that he has always remained a dear friend to me. In Meyer was a fortuitous combination of grace and unusual intellectual gift. In addition, there was a natural amiability and a great talent to adapt himself to the surroundings, so that people quickly found sympathy in him.

His brother Richard dedicated a detailed biography to him and a further remembrance through the publication of his letters. Nevertheless, it seems to me not only justified, but also a form of duty as a friend, to give a characterization of him here. He was a quick thinker and had a thorough and extensive command of information, a consequence of his wide reading and excellent memory. Ideas gushed out from him like a fresh and inexhaustible mineral spring, without him losing the [capacity] for healthy critical thought. This explains his unusual success in experimental chemistry, where a creative imagination must be combined with a sober selection of solvable problems and the simplest experimental requirements. It was very interesting to hear him talk about colleagues, whose suggestions he gladly acknowledged and whose weaknesses he illuminated frankly, without any malice, [but rather] more in humor.

In addition, there was a bit of artist in him, with heartfelt joy in music, declamation, theater, and poetry in the widest sense. This erupted from him spontaneously and with natural charm from time to time, so that, chemically speaking, it seemed like a transmutation from natural researcher to artist. Unfortunately, at that time his nervous system was already shaken from excessive work, perhaps also from too generous enjoyment of life's pleasures, so that 1 year later in Zürich he collapsed and had to take vacation. It was around this same time that, for other reasons, I had to give up laboratory work. Luckily, we both experienced a form of rebirth, which, to be sure, for Meyer lasted only about 12 years.

In October 83 we met together with Baeyer for a short time in Munich and experienced the joyful surprise, over the course of a cozy hour of conversation, that Baeyer offered us close friendship. We both honored the excellent and dear teacher and had now earned the right to be allowed to call him "friend" in an intimate sense. As far as I know, we remained the only chemists who could boast of this privilege.

In our circle of bachelors at Erlangen there was no more interesting occurrence than the engagement of one of our members. In the winter 1883/84 we experienced this three times, with Ludwig Knorr, Leo Gerlach, and the already 40-year-old medical doctor Kieselbach. They were naturally discharged from our circle with celebrations, and we [joyfully] took part in the weddings. The first, for Leo Gerlach, took place during the Easter vacation 84 in Nürnberg, because the bride belonged to the old-established family Seitz there. It was held with the great, somewhat stiff pomp of the old city of art. The second marriage was more modest in outward proportions, particularly in the number of participants, but arranged with fine artistic taste. It took place in the house of the director of the Academy of the Arts Piloty in Munich. At the celebratory meal, the bride was so placed that [her] splendid blond hair, which had served as a model for the father in the ideal figure of Thusnelda in the well-known picture "The Triumphal Procession of the Germans," was illuminated solely by the sun and therefore surrounded by glory after a fashion. My neighbors at this wedding were the bride's sister Johanna, a [woman] distinguished by beauty and the later Mrs. von Hefner-Alteneck, and the elegant, glib baroness L. von Hornstein, the later second Mrs. von Lenbach.

Because of my friendly relations with Knorr, I had to give a speech to the newly-weds and relay to them the best wishes of our young Erlangen society, together with a large album of photographs. I had naturally pondered the speech in advance and thought out an introduction with the names Knorr and Piloty tied together. It was a comparison of ships, and now there was the coincidence, that just before me on the table stood a splendid sailing ship [model] finished in silver, from which I naturally proceeded. The speech thereby assumed an entirely improvisatory flow and at the end, the [bride's] father, whose [own] speech had gone quite sour, made it clear to me that we professors were far above artists [when it came to] chattering.

[After] these two weddings I came home ill, the result of the 2-year work with phosphorus pentachloride in the poorly-ventilated private laboratory and made worse by an acute cold. Without [stopping] to care for myself, I was soon traveling again to visit my father and brother-in-law at Uerdingen. Here I contracted, probably while hunting, a small injury of the intestine. I turned for treatment to the surgeon Professor Bardenheuer at the Cologne City Hospital. I was operated on and had to lie in bed 14 days. Unfortunately, a high fever set in, and under these unfavorable circumstances my bronchial infection developed into a [severe] cough. After being released from the hospital, instead of going south, I foolishly went to Euskirchen, took part in the hunt, and exposed myself to a new illness. It didn't help when [next came] a 14-day stay in Wiesbaden, where I [drank heavily] at bars in the evenings. The result was that my illness assumed a chronic form. Most heavily damaged were the nose, throat and trachea, and I was particularly unlucky that the nasal infection fully suspended my previously so sensitive sense of smell, and for nearly ½ year I had no sense of smell. Some blame for this perhaps can be assigned to the experiments involving smell, to which I will return later.

In spite of the illness, I held lecture and laboratory classes in the summer, because I believed that in the autumn vacation the illness could be cured. But I had still not assumed the proper lifestyle. I was passionately devoted to smoking and could not give it up, and perhaps I also drank more wine at that time than was good. Then in August, I allowed myself to be persuaded, by Fleischer and Penzoldt, to accompany them to Pontresina in Engadin [Switzerland.] The journey began already with an accident, which might have had quite bad consequences. In order to avoid the overcrowded post [coach,] Fleischer and I decided to rent a private wagon without informing ourselves about the characteristics of the horses and coachman. After several hours of tolerable driving, we met an Italian organ grinder who had put a brightly colored cover on his hurdy-gurdy. One horse started at this, the coachman lost control of the animals, and after the poor railing had broken through, we tumbled off the road, perhaps 5 m down the slope, luckily onto a meadow. I had seen the accident coming, had stood up and wanted to jump from the wagon. It was too late, though, and I flew in a wide arc from the vehicle into the meadow. Never in my life have I sprung back up from the ground so quickly, for I feared that the wagon would follow. But it had in the meantime turned over completely and remained lying, heavily damaged. Fleischer had also flown out [of the vehicle] and had somewhat severely sprained an arm. Curiously, the guilty parties, that is, the coachman and the horses, remained entirely uninjured.

The accident had been observed from a distance of perhaps 500 m by the guests of a small sulfur bath Alvaneu. When we turned in there, in order to recover from the fright with a midday meal, the table company would not believe that men who had just undergone a truly life-threatening [experience] could want to eat. I had now lost interest in traveling by wagon. We therefore sent our luggage ahead with the post [coach] and proceeded the rest of the way to Pontresina on foot. I remained here only a few weeks, because the stay in the dry and, at night, cold air; the small tours of the glacier; and the nightly lingering for hours in a smoky beer bar were only damaging to my condition.

I preferred, therefore, to go to the Flims health resort at lower [altitude] in Graubünden, where I met my friend Königs and spent several enjoyable weeks. Here I got to know Kappler, the president of the Swiss school advisory [board,] in rather curious fashion. In the beer house connected to the hotel, a company of older Swiss gentlemen gathered every evening, and they enjoyed [playing] "Jas," a popular card game in Switzerland. In this small circle, an old gentleman so distinguished himself through vivacity, quirky remarks, and powerful wit, that we enquired as to his name. It was Mr. Kappler, whose reputation was known by all docents of science in Germany. Our curiosity was revealing to the old gentleman. He made inquiries, and when we sat at table the next day, he sent the waiter to me with the question whether I were the Otto or the Emil, for he was excellently informed about the young natural researchers in Germany. We then came into personal contact, and he immediately expressed the wish that I assume the professorship of chemistry at the polytechnic in Zürich, as Victor Meyer would be moving to Göttingen at the end of the next winter semester. When I replied that I was ill at the moment and I would first need to recover, he [dismissed this concern] considering my healthy appearance. Several weeks later, after making a report to his colleagues on the board, he repeated his offer in a letter. The offer was very tempting, as a splendid new institute for chemistry would be built, the plans for which Victor Meyer and Lunge, in consultation with an excellent architect, had already completed. It would also have had the attraction for me to become Meyer's successor, but I was still unsure whether in my condition I would be able to cope with the strains of the Zürich professorship. Meyer, after all, had collapsed in the end and had lodged spirited complaints about the exhausting life and activity in Zurich. So, I declined once again.

On the return journey from Flims, I was again threatened with the danger of a wagon accident, for as Königs and I drove away toward Chur in a carriage-and-pair, both horses immediately collapsed on the smooth pavement in front of the hotel.

Luckily, we remained in the wagon unhurt, but my aversion to travel by wagon was only strengthened by the incident.

I spent the rest of the vacation in south Tyrol, Brixen and Meran, where because of the heat and the dust my bronchial condition did not fully heal. Consequently, on a short stay in Munich, I again contracted an acute illness and returned to Erlangen sicker than when I had left in August. I therefore came to the conclusion that a longer, more serious cure would be necessary, and I took an extended leave of absence, which the ministry in Munich approved in a friendly manner. In order not to leave the university neglected, I proposed to the faculty to hire my cousin Otto Fischer, who was *Privatdozent* of chemistry in Munich, as provisional substitute for the time of [my] illness. This occurred and led to the consequence, that a year later, when I moved to Würzburg, [he] became my [permanent] successor. Not only did the cousin assume my lectures and leadership in the laboratory, but even the apartment, including the housekeeper, for the time of my leave. I remained in Erlangen until the end of November, in order to introduce him to all the business, and during this time I had the especial pleasure of seeing Mr. Kappler from Zurich again.

After I declined [the position in Zürich, Kappler,] in spite of his age and poor eyes, decided to make a tour of Germany in order to get to know the young docents of chemistry. Accompanied by his daughter, he also appeared in Erlangen in order to make the acquaintance of my cousin and to visit his lectures. I invited him to table, and after we had passed an enjoyable time at table and he believed me to be in a good mood, he made a final attempt to win me over. He maintained that I was by no means so sick, he would calmly take the risk with me, and then he explained in detail, with astonishing eloquence, the advantages of Zürich, and he stressed the amenities offered there to a bachelor due to the freedom of mores. He soon saw, however, that he could make no inroads with me, so thereafter he limited himself to recounting a series of interesting incidents from his life, mixed with delightful anecdotes. He was a superbly educated, very clever man, outfitted with all the good characteristics of the Swiss, and he spared no pain to recruit the best possible instructors for his beloved polytechnic. The prosperity of this school at that time is surely due in considerable part to the work of Kappler. Four years later, when I notified him from Würzburg of my engagement, he sent me a letter which was just as interesting as it was charming.[2]

The only thing he regretted was that the engagement did not take place in Zürich. Then came a long explanation concerning the efforts of the Swiss to accomplish comparatively more in the area of teaching than the large European states, who devote so much intellectual power to politics and the military. When he took leave from me in Erlangen, he said he would no longer trouble himself with bad advice from others and would simply follow his nose in order to [find] the best possible successor for Meyer. His choice then fell to Hantzsch.

At the beginning of December, I left Erlangen, and as an extended stay at a health spa had become abhorrent to me, I first went to my brother-in-law Arthur Dilthey in Rheydt. His house was comfortably furnished and provided with central heating, and he had invited me to Rheydt for a cure at the winter resort in a manner that was

---

[2] Translator's note: Fischer became engaged three years later, in 1887, not four years later.

equally engaging and jesting. Here I spent three enjoyable months. The days were used for taking long walks, and in the evenings, supposedly to conserve the voice, I played skat with the brother-in-law and August Fischer. But we made such comical stupid [mistakes] that we couldn't stop laughing. Also, we assiduously consumed wine, so this was not exactly the remedy that was needed to cure my illness. But it also wasn't getting worse, and my frame of mind had improved enormously in the merry Rhenish circle. Naturally I also frequently met with my other brothers-in-law and the sisters, and the old aunt "Lisettchen," [who] likewise did not fail to invite the nephew from time to time.

During the week of Christmas, a small fire occurred at my brother-in-law's house, which might have had quite unpleasant consequences. The central heating had dried out the air, and the Christmas tree likewise had dried out. When the [candles on the] tree were relit at the wish of the children, the resin-rich needles caught fire, and in a short time the entire tree was ablaze. For me personally this was not an unusual spectacle, for in the laboratory one often experiences such quick fires. But for the family of my brother-in-law, particularly the children, [the fire] made an entirely paralyzing impression, and the children's governess so lost her composure, that she wanted to run directly into the fire to extinguish it. There was nothing for me to do but to send her and the children out and then issue the command: "let it burn out quietly." In half-an-hour this had happened, a few drapes were burned, a few pictures and carpets damaged. We contained the fire with a couple of buckets of water and the event was over. Since then, I have always warned against relighting Christmas trees in houses with central heating after [the tree] has been standing several days.

At the exchange of presents an amusing [interaction] occurred, which is characteristic for the artistic sense of the children. My brother-in-law had received a plaster cast of the bust of Venus di Milo as a present, which, due to the severed limbs, was criticized by the children in not-too-friendly [manner.] Suddenly little Elsa raised the question: "What are those bumps on the breast?" whereupon the younger six-year-old Alfred answered her: "How stupid, Elsa, those are furuncles."

In February 1885 I decided to spend the spring at the Mediterranean. My friend Victor Meyer, who in the meantime had suffered from a nervous breakdown and neuralgia, advised me to go to Ajaccio in Corsica to the Swiss Manor of Mrs. Dr. Müller. Never have I received better advice pertaining to a health resort. [I took] the journey via Paris and Marseilles. One evening in Paris, on a small stroll along the boulevard, I lost my gold watch and noticed [the loss] only after taking about 20 steps further. Naturally, I immediately turned around and had the good fortune to find the watch still on the ground, even though many people had passed the spot.

In Marseilles I arrived with a large letter of credit from a Cologne banking house, which was made out to a somewhat large sum, because I had the intention to add on a sea journey to Brazil following the stay in Corsica. When [I arrived] with this letter at the office building [of the house] on which the letter was drawn and wanted to withdraw several thousand francs, I was informed that the till did not have enough [cash.] There was much laughter, and I had to wait until the money was fetched from the bank. Marseilles, which I already knew, renewed its great appeal to me because of its splendid setting and the handsome harbor buildings.

The sea journey to Ajaccio, which lasted about 20 h, is still a fond memory. The French were most polite at table. There were also a number of prisoners on board who were to serve a long prison sentence in Corsica. They entertained themselves side by side in the mild, starry night on the entirely calm sea with songs and frisky jokes. The entire company was influenced by this merriment, and I also had the agreeable feeling that the mild, moist air of the Mediterranean would be the proper means to cure the diseased mucous membranes of my breathing passages.

Early in the morning we already saw the snowy mountains of Corsica beckoning, and at the arrival in Ajaccio we were presented with such a splendid view, the likes of which not many other places on the Mediterranean offer.

We were received at the guest house in a most friendly manner, as [business] had still been quite weak due the cholera epidemic the previous summer.

In the months of April and May, Corsica has climatic conditions so good, like few [other] places on the Mediterranean, very much sun and a cool, fresh sea breeze at midday; in the evenings at sundown likewise a short cool-off period and then steady temperature until late at night. As [I] stayed outside the city, [I] also remained protected from dust, which is so troublesome for people on the Riviera, for in Corsica there are roads, but no horse-drawn vehicles. Under these favorable conditions, my bronchial condition was healed in 8 weeks. There remained, however, a tendency for acute relapses, which for several years necessitated that I avoid, as much as possible, catching cold and also the hazardous gases of the laboratory. It was only after 33 years that I again suffered an influenza-like bronchial infection that would not heal and after six weeks led to an inflammation of the lungs. This double illness was the reason, in April and May 1918, for a six-week cure in Locarno on Lake Maggiore. I used the involuntary leisure time to write down the first part of this memoir.

Corsica is a wild mountainous country, whose tallest peaks, the Monte d'Oro and Monte Rotondo, always have snow and even small glaciers. The massif consists of granite, broken through here on the Mediterranean [coast] by the predominant chalk. Something of the wildness of nature shifted to the population, for they are well-known as bold warriors and gladly boast of being the people of Napoleon Bonaparte. For centuries they have practiced the blood feud, and still in my time there was a secluded village in the region that was populated almost exclusively by such murderers, who had taken refuge here and not infrequently engaged the French rural police in armed resistance. The barbaric customs of the countryside are precisely portrayed in the well-known book by Gregorovius, and he who wants to get to know [these customs] in a more charming portrayal should read the superb novella "Colomba" by Prosper Merimée.

Respect for family honor has seemed to be primary duty to Corsicans for centuries. We experienced an excellent example of this. In a neighboring hotel, an attorney from Zürich tried to seduce a Corsican servant girl. The outcome was negative. But after a short time, two relatives of the girl appeared, farmers from the surroundings, naturally armed to the teeth, and asked the attorney for an interview. He was clever enough to declare that he had made the girl an honorably intended marriage proposal. This had a calming [effect,] and the two men departed after issuing the strong advice that he should not bother the girl again. According to the declaration of experts, these men would have shot down the attorney in the hotel, if he had not delivered a statement sufficient to satisfy the family honor.

There was much too much shooting in Corsica, for hunting was free, and every grown man felt obliged to take advantage of it. On longer walks one had to take precautions not to be shot at by careless hunters. The Corsicans seem to have little interest in peaceful work, for the fields were neglected throughout, and where diligent workers were at work, one could be sure that they were Italians recruited from the mainland.

My infection had healed, and as I received the news that because of the death of Kolbe, a shuffling of the professorships of chemistry would next occur, I gave up the trip to South America and returned via Switzerland to Erlangen. Along the way I visited Carl Graebe in Geneva, who showed me the new, expensively built university laboratory. Later, during vacations, we met more frequently in Switzerland or in the Black Forest, and I will have more to say later about this excellent man.

My leave ran until the autumn, so I spent only a few weeks in Erlangen in order to greet my cousin Otto and other friends. I then went for an extended stay to Baden-weiler in the Black Forest. On the way I stopped for a short rest in Frankfurt and visited the Höchst Dyeworks, my friend Vongerichten who worked there, and above all Dr. Lucius, partner of the firm, who had previously called on me for a visit in Erlangen. Not only was I received at the factory but also in the family Lucius in a most friendly manner. I remained in communication with [Lucius] until the Berlin time.

In Badenweiler, with its splendid surroundings, I continued the life of idleness and luxury from Corsica for perhaps another 5 weeks, until one day a telegram arrived from the Würzburg professor Semper, who invited me for an interview in Heidelberg. It was already known to me that J. Wislicenus had been called to Leipzig as Kolbe's successor and that the professorship of chemistry at Würzburg had to be filled again. I thought it improbable that one would consider me since I was on leave due to illness, and these things are always greatly exaggerated by rumor. In fact, the faculty in Würzburg had declined [my candidacy] and proposed O. Wallach to the ministry in Munich. However, before the decision was made, Miss Bertha Strecker of Munich, the sister of Mrs. Leube, learned in Erlangen that I was healthy again and brought this news to Würzburg. This presented Professor Semper, a member of the faculty, with the possibility for the consideration of my appointment again, and for that purpose he arranged the meeting.

[The meeting] took place at the Hotel Schlieder in Heidelberg. I soon noticed that it played out as a test of my health condition, for which, apparently, Semper as a zoologist was particularly well qualified. Later, when the affair became rumor in Würzburg, it was said that they had had me examined by a veterinarian. Enough, Semper made the suggestion to undertake a stroll to the castle. Although he was much older than I, he deliberately set a rather quick pace, with the result that he arrived at the top quite breathless, while I, being used to climbing mountains, felt quite comfortable with the pace. Then came the second test. Semper suggested to drink a bottle of champagne. This was also not disagreeable for me, for the enjoyment of wine was among my habits. The result of this breakfast was then also, as one might have expected, a slight drunkenness of the older gentleman [but not] of the younger colleague. The exam was passed.

Semper traveled back to Würzburg and declared to his faculty colleagues "the Fischer is an entirely strong, capable man who will outlive all of us," which [proved] to be correct. Consequently, a new recommendation of the faculty went to Munich, and perhaps a month later I actually received the call to Würzburg from the ministry. I felt bad for Wallach, for he had been informed of the first recommendation of the faculty. But I could not backtrack because of this, and as the change to Würzburg was extraordinarily agreeable to me, I gladly accepted the offer.

Before [the move] I made a stay of several weeks at [Bad] Homburg, from which two occurrences have remained in memory.

First there was another wagon accident. I visited Mrs. Dr. von Brüning, née Spindler from Berlin, whom I knew well through her husband [when he was alive,] at her splendid estate near Homburg, and she invited me on a drive. I tried to decline, for the reason that wagon drives present much danger to me. But she did not consider this to be a valid objection, because horses and coachman had proved reliable to her through many years of service. We therefore drove, accompanied by her son, to the nearby Saalburg in the Taunus. As we merrily set off on the return journey, we were on the level road, when suddenly the coachman gave a cry and fell senseless from the box. The reins naturally scraped along the ground. Luckily, the old horses calmly trotted on, until the young Brüning climbed from the wagon up to the box and could grab the reins. So, the incident passed without any harm to us. The ill coachman was placed in the wagon and driven home. We followed on foot, and Mrs. von Brüning was somewhat shaken by the prompt realization of my accident prediction. Consequently, in the future she always had another servant sit in the box next to the coachman.

Coincidentally, the family Meister had also obtained a summer apartment in Homburg. On a visit there, I was invited for company in the evening, at which the Princess Bismarck and her son Herbert took part. It was naturally very interesting to get to know the next-of-kin of the great chancellor of the empire. She seemed a simply natural person. This was especially evident from the princess, who spoke of herself in an almost gamine manner. Mrs. Meister was the sister of Mrs. Lucius, and both were again daughters of an artist painter from Frankfurt-on-the-Main, in whose house Bismarck had mingled when he was the Prussian representative at the *Bundesrat* in Frankfurt. It was from there that the friendship stemmed between the two ladies and the Bismarck family, and that had brought me the unexpected honor to meet with the princess and her son.

Chapter 7
Würzburg

In the meantime, the appointment at Würzburg had come to pass. Therefore, I went to Erlangen to arrange the move. Shortly thereafter I paid a visit to J. Wislicenus in Würzburg in order to inform myself of the conditions there, in particular, also, about the equipment of the institute. We already knew each other from conferences of natural researchers and from festive events of the universities at Würzburg and Erlangen.

[Wislicenus] was a pleasant individual, with a very dignified outer appearance and engaging personality. He was patriarch of a large family. He was generally respected as a benevolent teacher and distinguished character. The University of Würzburg had twice elected him Rector, and as such he led the 300-year celebration of the school in exemplary fashion. His life's work is portrayed in a biography penned by Beckmann, and no one will deny the merit he earned through his work on lactic acid, the synthesis of acetoacetic ester, and later the many suggestions in the field of stereochemistry. However, as a chemist and natural researcher, he represents an entirely different type than Baeyer, Hofmann or Liebig and Wöhler. He relied on preconceived theoretical opinions to test by an experiment, rather than to search for unexpected processes by following empirical results. He probably would not have been able to manage investigations where great experimental difficulties had to be overcome. This was evident from the somewhat sorry equipment of the Würzburg institute, which had been built by Scherer 20 years earlier without considering the needs of our rapidly progressing science. The first thing I did in Würzburg was to make the suggestion to improve the insufficient ventilation through installation of new ducts. Wislicenus supported me in the most obliging manner and warmly endorsed my request with the director of the administration, Professor Risch, who had fairly large resources from the income of the school at his disposal. The negotiations resulted in an approval of 6000 marks. Risch insisted on a solemn promise from me that I would never again call upon university resources. I laughingly agreed, with the remark that I would

D. M. Behrman and E. J. Behrman, *Emil Fischer's "From My Life"*, Springer Biographies, https://doi.org/10.1007/978-3-031-05156-2_7

break [the promise] at the next opportunity. This occurred already after a few years, and I must add to the credit of the colleague Risch, that he always had an open hand for my wishes.

The small new installations in the institute were carried out during the vacation, so that by the end of October the work was finished. I had thought out a very simple system for the installation of the ducts. Clay pipes, such as those used for toilets, were laid on the walls fastened with iron clamps, then led through the roof and protected there from the wind by a suitable superstructure. Below, in the workroom, an elbow was installed to arrest the fall of dirt from the pipes. Around the lower opening of the pipe the actual fume hood was built from wood and glass. A pilot flame in the pipe provides the necessary draft. This simple form of the fume hood [should be] installed in every building and is therefore highly recommended wherever provisional laboratories are established in existing structures.

J. Tafel and the family Knorr, who now had a son, moved from Erlangen [with me.] No suitable apartment was available for [the Knorrs,] so it was my pleasure to accommodate them in the large service apartment of the institute until they found quarters [off-campus] in the spring.

The old assistants of the institute had followed J. Wislicenus to Leipzig. Against that, he had left behind for me his son Wilhelm, to whom I was glad to entrust an assistantship, and he later became a dear friend to me. He was also able to find quarters at the institute.

Knorr, who had already qualified for a position as lecturer at Erlangen, was accepted by the faculty as *Privatdozent* at my request without further [ado.] He was also entrusted by me with leadership of the analytical division in the institute. The only large work room in the building was reserved for the [analytical division] while the organic division had to make do with a small annex and the cellar underneath. Everything [there] was quite shabby and also impractically arranged. The lecture hall was fairly roomy, because a large number of medical [students] had been expected; even so, the equipment left much to be desired.

My private laboratory consisted of a single simple living room, and as space for weighing and optical investigations we used the consultation room next door. Nevertheless, here the experimentally quite difficult work on sugars was carried out. I began, though, with the development of the synthesis of indole derivatives from phenylhydrazones, [work] which started in Erlangen. Most of the doctoral students took part, whereby the institute immediately took on a bad "odor." The main stinker, skatole, that we made in large quantities, was introduced by the lab workers into the guest houses and wine houses of the city and was the cause of many complaints. I experienced a drastic example of the persistence of its odor.

In the Easter vacation 1886 I undertook a second trip to Corsica and took along the full-cloth blazer that I had worn in the laboratory during the winter and [intended] for service on small mountain tours on the island. When my suitcase was opened by customs at the French border, the official refused, with a gesture of indignation, to look through it further, and demanded that the case be closed immediately; for a strong odor of skatole had arisen from the case and had apparently given the official the impression that the case must contain underwear very heavily soiled with excrement. It went even worse at the Swiss Manor in Ajaccio. I had hung up my clothes in the room unawares. But after several days the hostess appeared and served up the urgent question: "In God's name, what are you doing in your room? The neighbors are complaining about a bad smell coming from here." Now I immediately knew where to lay the blame, and the embarrassing blazer was hung for 14 days in the open [air] in the Corsican sun. When I packed up and returned to Germany, the smell was greatly reduced, to be sure, but by no means had it disappeared.

From the outset in Würzburg, I had the good fortune to draw a succession of splendid doctoral students. I mention first A. Schlieper, son of the cotton printer Adolf Schlieper in Elberfeld, who 30 years earlier, through his observations in the uric acid group, gave Adolf Baeyer cause to occupy himself with these materials. The son became his successor, and he now stands, if I am not mistaken, at the summit of the firm Baum & Co. C. Steche, who came to Würzburg at the same time, had an equally laudable career. He is now the leader of the large perfume firm Heine & Co. in Leipzig. Then Ahrheidt, later manager of the Aniline and Soda factory of Baden. Friedrich Ach, from Würzburg, a chemist equally diligent and gifted, was an ardent member of a dueling society and contributed by seeing that his society brothers did not ignore the lectures. After he finished his studies, he entered the alkaloid factory of C. F. Böhringer & Sons in Waldhof near Mannheim and there, as leader of the scientific laboratory, took a large part in the development of the synthesis of caffeine. I will report more on this in the [chapter] on the Berlin time. Unfortunately, he died early from psoriasis and nephritis.

Further, Jacob Meyer, who later independently discovered tannigen, for several years now at the Berlin institute, is working on a variety of problems and provides me help in the administration of the institute.

In the summer 86 I turned again to the osazones of sugars, already discovered in Erlangen, and I succeeded in obtaining isoglucosamine with support from W. Wislicenus, who at that time was my private assistant. Soon thereafter, after the work on indole had been concluded, [we] began the syntheses in the sugar group, with which I chiefly occupied myself until the end of the Würzburg period. The number of my coworkers here was so large that I cannot mention them all, but I feel obligated to name three in particular. First, J. Tafel, who took part in the oxidation of polyhydric alcohols and the transformation of glycerose or acrolein dibromide to the first synthetic sugar with 6 carbon atoms, the acrose. That was very arduous work and, in part, quite hazardous to the health [of the researcher.]

In order to obtain sufficient quantities of acrose in the osazone form, we [spent] several weeks together at the dyeworks Meister Lucius & Brüning in Höchst-on-the-Main preparing acrolein and its dibromide with the help of the large resources of the factory. The large vessel from which the acrolein was distilled was located outside. The operation, which had to be repeated several times, was carried out on windy days in order to dissipate the terrible smell of the acrolein. Tafel once wandered into a cloud of acrolein by mistake and got so severe a nosebleed, that I feared for his health. Somewhat later, as ever more difficulties in the sugar work piled up and a lucky solution to the problem seemed possible only in the distant [future,] Tafel turned to other projects, since for his intended scientific career, he also needed independent results. A succession of other gentlemen stepped into his place, among whom I must name Josef Hirschberger, now in Brooklyn, and Heller, now in Leipzig. The Englishman Passmore, who is now a respected commercial chemist in London, and Lorenz Ach from Würzburg, about whom more later, took part in the special syntheses of carbon-rich sugars through the cyanide process.

The older assistants were occupied with their own problems independent of me. Knorr had had the great practical success with Antipyrin in Erlangen and scientifically [speaking,] discovered not only pyrazole but also the elegant synthesis of pyrrole derivatives from 1,4-diketones. These findings were developed to the greatest extent in Würzburg, including the synthesis of morpholin, from which the researcher assumed that it contained the nitrogen ring of morphine.

Wilhelm Wislicenus made another strike just as fortunate, with the extension of the acetoacetic ester synthesis to other esters, for example, the oxalate ester. Were it not for L. Claisen's observations at the same time, he would surely have conquered the entire large field of condensation of esters with one another or with ketones and aldehydes. Another very pretty observation of his was the formation of sodium azide from nitrous oxide and sodium amide. About this same time Tafel found the reduction of hydrazones to amines using sodium amalgam, and he would surely have also extended this reaction to oximes had not Goldschmidt quickly carried out such experiments and published soon after Tafel's results [were known.]

Among the Würzburg pupils, a special mention is earned by Oscar Piloty. After he had no success with the association exam in Munich, he came to Würzburg in a somewhat dejected state. But here he quickly regained his self-confidence after he was able to demonstrate his experimental talent in his doctoral work on the carbon-rich sugar rhamnose. He then successfully completed his doctoral degree and afterward took part in the quite difficult transformation of saccharic acid to glucuronic acid. At that point he was allowed to marry Baeyer's only daughter Eugenie, with whom he had long been as one. The young couple came to Würzburg, but an unfortunate occurrence during the honeymoon landed the groom in jail for a short time, for he had forgotten about a [required] military inspection. To my great pleasure, he decided to move to Berlin with me, to work there nearly seven years as assistant and later as independent leader of the analytical division. But when the opportunity arose to receive a professorship at the Munich laboratory, the old yearning for the Bavarian homeland irresistibly pulled both him and the wife there, although he could have had the same position one quarter of a year later at the new Berlin institute, perhaps under better circumstances.

Piloty's fine scientific accomplishments, which should be related in a detailed necrologue by other colleagues, have thoroughly confirmed my original opinion of his experimental talent, and I have taken even more satisfaction, as his father-in-law did not at first share this opinion.

At the start of the unholy war, Piloty was already beyond the age of military duty. Nevertheless, he volunteered out of patriotic enthusiasm, and I was able to facilitate his promotion to lieutenant of the empire with a recommendation to the Bavarian military authority. He fell in the Battle of the Somme[1] at the head of a machine gun unit. He was deeply mourned, not only by his family, but also by friends and colleagues, who expected many more beautiful scientific results from him.

In the last year of the Würzburg period a young Dane Dr. Fogh came there, in order to carry out a thermochemical project, mainly on the compounds of the sugar group. He had previously become familiar with thermochemical methods with M. Berthelot in Paris, and he had also brought a set of instruments with him from there. He was lacking only the bomb calorimeter, but he could travel to Paris during the vacation and carry out the calorimetric experiments there. The work appeared in the "*Comptes rendus*". Dr. Fogh went likewise with me to Berlin and was for a time assistant in the inorganic division there. But for reasons of health, he soon had to take a leave and finally returned to Copenhagen, from where he sent me an engagement notice. Since then, I have heard nothing more from him.

In the large experimental lecture in chemistry the majority of the audience are medical students, then come the apothecaries, and only in third place, the chemists. I have taken the trouble to become fair and keep the actual lecture as simple and popular as possible without endangering the scientific character. But afterward I often gave a supplement for the career chemists, wherein not only difficult theoretical questions, but also many special experimental methods were discussed. So far as I could tell, this form of the presentation was very welcome to the students. Unfortunately, it was very demanding for the docents.

The collegial interaction in Würzburg differed not insubstantially from the Erlangen customs through the free, for the most part totally informal expression of meaning and through the ingenuous manners. The pompous display of academic rank practiced in the Erlangen Senate as well as by the faculty was little to be felt in Würzburg, which struck me, as a Rhinelander, quite pleasantly.

---

[1] Translator's note: Piloty died on October 6, 1915 at the second battle of Champagne, near the Somme. The (first) Battle of the Somme was waged July–November 1916.

The faculty [at Würzburg,] just as at Munich, was divided into two sections. Our mathematical-scientific division included only 6 full professors, but under them were many persons distinguished by high scientific esteem. This was particularly true for the physicist Friedrich Kohlrausch and the botanist Julius Sachs. I soon entered into friendly relations with Kohlrausch, for he was an equally sensible and kindly colleague and was always ready to support reasonable wishes. I need not discuss here his excellent investigations on the electrical conductivity of solutions and his determination of physical constants, as well as the superb textbook "Practical Physics" published by him, as they are sufficiently well known to natural researchers. I took many a stroll with him on beautiful summer days, and I also found a friendly reception in his family, particularly on the part of his charming wife.

Sachs, the famous plant physiologist, was of an entirely different nature. He established a large school of botanists in Würzburg and attained prominence through a large textbook and the detailed history of botany. [His] withering judgment of the purely schematic system of Linnaeus was characteristic of his bold criticism. He liked to say that this system had come like a miasma over botany and destroyed the healthy seeds, already present, of a natural system for a long time. During my time Sachs was already an ill man with overstrained nervous system who could vent his moods in vehement manner. Nevertheless, I succeeded in establishing friendly relations with him, and I learned much in frequent visits to his institute which was useful in my own work in the physiological chemical area.

I remained more distant from the minerologist Fridolin Sandberger, likewise a meritorious scholar and a benevolent individual, but an oddball who still recognized the German Empire only grudgingly and also wanted to know nothing of modern chemistry. He expressed this drastically with the saying: "Chemistry is that which explodes and stinks."

The zoologist Karl Semper, who played such a large role in my appointment, was likewise at least as eccentric. The prime of his scientific [career] was also already awhile past. As a young man he had traveled widely for scientific purposes and wrote valuable books about the Philippines and the Palau islands, where he stayed for a year and also met his future wife. During my time he occupied himself mostly with the plan for a new building for the zoological institute. As the approval in Munich for the moneys ran into difficulties, he tried the most curious means to make clear to all people the necessity of the building. In the old institute, which was accommodated on the fourth floor of the university building, he installed large aquariums, which one day became leaky and flooded all the rooms below. There was nothing else to do but erect a new building on the Pleicher Ring. The opening ceremony was so impressively staged by Semper that surely the memory remains with every participant. At the end, he led the entire company, among whom were the leadership of the university and administration departments as well as a succession of ladies, into the hothouse, where plants and animals were assembled in colorful alternation, and which was provided with a comprehensive irrigation system. As the company enjoyed beer and little sausages in a festive mood, with the doors closed, the irrigation system suddenly came into operation and the entire assembly was drenched in a tropical rain.

Sometime later the culture minister, Dr. Müller, who had just entered his post, came to Würzburg for an inspection of the university. The individual scientific institutes had been notified of his visit; however, for whatever reason, no mention was made of the appointed time. The wait became too long for Semper, so he had ordered a breakfast of sausage and beer and was engaged in devouring it when the minister entered. The latter greeted the heartily-eating professor with the jovial words: "it seems that you are enjoying [your] food today," whereupon the prompt answer came: "If you had waited in vain so long for the minister, you would also have become hungry." This unreserved answer, distinguished not at all by deference, gave the minister cause to take his leave soon thereafter.

These are only a couple of [examples] of the many amusing stories that Semper delivered. On the other hand, he was a splendid man of worthy comments, full of amusing ideas and free from scholarly pedantry. One perceived with him the extraction from the family distinguished by superb artists, for example, the famous architect Gottfried Semper.

The mathematician Prym was another quirky character. He stemmed from a rich manufacturing family from Düren. He had studied mathematics, then [worked] at a bank in Vienna. From there he was called by Kappler to Zürich [to become] full professor of mathematics at the polytechnic. Although blessed with riches, he led quite a simple life with his family in Würzburg. He took his teaching profession very seriously. The lectures were most scrupulously prepared, and it occurred not infrequently that he had absent students picked up from the apartment and driven to the university in his wagon. As an associate with many industrial concerns, he was interested in chemistry and later also visited me in Berlin many times.

A second full professor of chemistry was not present at Würzburg. The institute for applied chemistry was led by Associate Professor L. Medicus, who had been Assistant not only to Wislicenus, but already to his predecessor A. Strecker, and therefore was the main person responsible for the chemical tradition in Würzburg. Great scientific interests never troubled him, but he was quite popular with the students because of his joviality, and he was a welcome member of our chemical circle because of his delightful sense of humor and his friendly outlook on life. After my departure he was made full professor and a few years ago also went already to his forbears. He took part in the military campaign 1870 as a Bavarian officer and was wounded [and] taken prisoner at Orleans. His descriptions of the poor treatment and hateful derision on the part of the French population exactly matches what one hears today about the treatment of German prisoners of war in France, although at that time the war ended quickly, and the discontent of the people was by no means whipped up [to the extent] that it is today.

The philological-historical branch of the faculty had few outstanding men aside from the classical philologists and the archaeologist Urlich. It was entirely different among the medical faculty, with whom I had many connections. Wilhelm Leube, who moved with us from Erlangen, became successor to Professor Gerhardt, who had been called to Berlin. [He was installed] at the internal clinic of the Julius Hospital, and he knew to uphold the great tradition of this position in every respect. Then there was the surgeon Maass, a talented operator, a stimulating teacher and excellent in company. Unfortunately, he died at quite an early age.

The senior faculty member was the anatomist Albert Kölliker, born Swiss. Through an abundance of worthy investigations, he enjoyed a fine reputation. It was due to his influence that anatomy apparently had the finest institute and played a prominent role in the education of medical students. In addition, he was an unusually handsome man, with a fine face, white curly hair, clever dark eyes, and a delicate, almost womanly skin color. He must earlier have become peeved with chemistry; for he was rumored to have said that he would let his most stupid son become a chemist, whereupon the spiteful citizens of Würzburg added that in fact he selected the right one. He was always friendly with me. His dissector was the Associate Professor Stöhr, a native of Würzburg, a very comfortable and engaging man, who also became Kölliker's successor.

The outpatient physician Geigel distinguished himself through eccentricity. [He was] a master at the composition of exquisite and slightly ironic testimonials, which the faculty always made use of when they wanted to dispute unreasonable demands from the ministry in Munich. He was head of a music group which consisted of professors or citizens of Würzburg, for example the brothers Stöhr and the pharmacologist Kunkel. During the autumn vacation this society resided in Ammerland on Lake Starnberger and gave small concerts daily that featured the horn.

At that time, the once-so-famous Scanzoni still functioned as obstetrician, [but] he soon died and was replaced by Hofmeier.

I must mention the ophthalmologist Julius Michel, an entirely peculiar [individual,] with whom I became good friends. He was a merry native of the Palatinate who had also been in Erlangen before my time. He took his medicinal trade and also the science very seriously, and as a physician as well as a scholar he enjoyed a good reputation. However, letting loose with mischievous wisecracks had become essential to him. In company he therefore was a stimulating character, and at Leube's hospitable house we not infrequently listened with pleasure to his jokes for hours. Even in serious moments he could not suppress a joke, and his lecture was somewhat peculiar in this connection. I want to cite just one example. One day he was testing a candidate of medicine by the name of Jerusalem. The result was unsatisfactory, and he now communicated the result to the crowd of the candidate's friends with the words: "Let Israel lament, Jerusalem has fallen." Of course, spiteful people later maintained that he had only allowed the candidate to fail in order to be able to produce this witticism. He was unmarried but led a splendidly managed household and gave small parties at which boisterous merriment prevailed.

Together with an assistant, I once had a small accident in which a filter bag burst and a strongly colored chemical preparation flew into the face. Not only were the eyes smeared but also irritated. There was nothing quicker to do than to seek help at the university clinic. Friend Michel immediately cleaned us carefully. However, after it was determined that the incident was minor, it gave him a special pleasure to write down in the book of the outpatient clinic a detailed record of this unusual case, wherein he gave an entire list of long chemical names to demonstrate his scientific importance.

Later, in Berlin, where he was appointed about 10 years after myself, I had to call on his help yet again, likewise with best success. At that time, I was troubled with mild but unrelenting headaches which [sometimes] led to nausea. The physicians whom I asked made quite apprehensive diagnoses concerning disturbance of the circulation and the like. However, I made the observation that the condition was the worst when I had been reading for hours the day before. So, I made the conjecture, that the eyeglass I was using, which I had once selected myself at the optician and had worn for over twenty years, was to blame. I therefore went to friend Michel. After he had examined the eyes and the glass, he summed up his judgment in the following words: "You earned these headaches long ago. People say you are otherwise not so stupid." He then prescribed the correct eyeglasses, and the headaches, which had bothered me for weeks, vanished. Never has a remedy for me met with such brilliant and quick success.

Among the jurists, aside from the director of administration Risch, already mentioned, the pandectist Burkhardt and the national economist Schanz have remained in memory. I also gladly remember several of the theologians. They were Catholic gentlemen, but by no means of the type of the so-called firebrand chaplains, but rather in part very refined, tactful, and jovial gentlemen. They called [their] senior faculty member the "bishopmaker." I often walked along the city's glacis, which offered such pretty strolls, with this [man,] and for hours we freely discussed church, state, and society. He had been educated in Rome at the *Collegium German-icum* and had written a book on church and state that was well worth reading. I heard a great deal from him about the institutions of the Catholic Church, particularly about practical ministry, which must have interested me to the highest degree. Whether this discourse earned me confidence in the Catholic circles despite my Protestant origin, I can't say. One day, however, I experienced the surprise that about 25 Catholic theologians attended my lecture on inorganic chemistry as registered [students.] When I asked the reason that these gentlemen had come, the answer was that the Bishop of Speyer had ordered it. Probably this ecclesiastical dignitary had the wish that the active clergy in his diocese should have certain chemical knowledge, in order better to communicate with the workers of the numerous chemical factories in the region. The religion teacher of the Catholic *Gymnasium* was also [in attendance,] and he declared to me one day that he could use knowledge of the modern chemistry superbly in religious instruction. From this I came to the conclusion that natural science and religion need not be enemies in principle.

In Würzburg there are two scientific societies in which I gladly took part. The older and more general [one] had the name Physical-Medical Society; its members were mostly professors and docents. The newer Chemical Society had only [recently] been founded by J. Wislicenus. Naturally the professors and assistants of both chemical institutes belonged, but the majority of members were actually older students. Original work was presented here, more frequently papers from the outside, and social life played a main role as well.

The zenith was the annual commemoration of the founding [of the society,] at which many previous members from afar and also many professors of the philosophical and medicinal faculty appeared. Usually, aside from witty after-dinner remarks and speeches, there was also a [celebratory] play. Two have remained in my memory. In one the doctoral exam of a [student] was staged with very comical and drastic attacks against the well-known examination methods of individual professors, with the result that the deacon of the faculty, a philologist, indignantly left the room while the rest of the audience were immensely amused. The second production was an operetta "The Chemikado" with melodies from the Mikado and a very witty text by Dr. Reitzenstein. The city theater had lent the costumes and wigs. The Mikado was my dear nephew Alfred Mauritz, previously mentioned, who was studying chemistry in Würzburg at the time. In later years I have experienced great friendship from him. He wore my laboratory suit and hat, spoke the lower Rhine dialect, and stood out that much more as the only European in the Japanese company. I have a photograph of the players, one glance at which still brings me to laughter.

The Chemical Society adhered to its traditions after my time, and many a pretty play was produced by Dr. Reitzenstein. Unfortunately, due to lack of time, I was unable to participate in the annual celebrations [after I moved to] Berlin. However, for the 25th-year anniversary I felt obliged to clothe my regrets in rhyme, and as this is one of the few poems I have composed, I want to quote the telegram here.

[to the] Chemical Society of Würzburg

As merry the festival

I wish you the best of all.

Flourish always so brightly

And remain ever spritely.

Greet your delegations

And chemist relations

From the old poison disher

Aemilius Fischer

In response, the Society promptly sent a charming verse by Dr. Reitzenstein.

It was no wonder that gaiety and humor thrived in Würzburg. The friendly city, with the splendid castle, the lovely river, the beautiful setting on the glacis, and the [grape]vine-covered mountains, the cozy lower-Franconian population and the old tradition of the episcopal staff were well suited to strengthen the already merry mood of the academic society. Discourse among the professors and also with the students was easy and comfortable and only occasionally, for example, during exams, assumed a more serious form. Nevertheless, a good spirit prevailed among the students, for overall in Würzburg a considerable [amount of] work was accomplished, and in the chemical laboratory one could not complain about a lack of diligence. The majority of the students consisted of north Germans. The same was true of the professors, and the particularism that the Bavarian government was often accused of in the appointment of professors was not apparent in Würzburg. A Protestant was even appointed as teacher of canon law at the university of the old bishop city, where only Catholic theologians [were on the] faculty! An intervention into professional obligations indeed was tried from time to time by the ultramontane majority [in the legislature] but for the most part successfully rejected by Dr. Lutz, the culture minister at that time.

Intercourse among families of the various faculty was also practiced in a charming manner in Würzburg. During the first winter, the Erlangen company, that is, the married couples Leube and Knorr as well as my trifling [self,] was the object of celebratory greeting at the typical evening meal, and we reciprocated alternately. This first after-dinner speech is for the academic community not infrequently the critical measure placed on new members. I had to speak at Kohlrauschs and pondered a humorous toast. In it, electricity played a role toward the end. I compared the individual ladies with the recently invented incandescent bulbs, reserving the arc lamp for the housewife. Now everyone would believe that I had tangled myself in a spider's web and that only through a violent rending would I be able to free myself. But luckily it occurred to me that from the arc lamp to the sun only a short leap is needed rhetorically, and with that I had obtained my poetic portrayal to glorify the housewife in a dignified [manner] and won the approval of the company to a cheer.

I had ongoing company among family only with the Knorrs and in particular with the dear Leube couple. We came together there at least once-a-week. Usually, Michel and a few other friends were there, and we passed merry hours together over a simple evening meal. Leube was a splendid companion, clever, well-educated scientifically, and stocked with the rich experiences of a successful medical doctor. He knew a large number of people who called on him for medical advice, spoke well, and composed absolutely delightful poems for the occasion.

His dear wife Natalie felt chemically related to me, for she was, as I mentioned earlier, the daughter of Adolf Strecker, who died as Professor of Chemistry in Würzburg. She became engaged with her Wilhelm as 18-year-old girl at the chemical institute. In this way, I fortuitously became Strecker's scientific heir, for example with the hydrazines and caffeine. Shortly after my marriage, the Leubes offered me the friendship of intimate [address,] and we now still remain in communication by letter, the Leubes spending a comfortable retirement in Stuttgart. The oldest daughter Lilly married the gynecologist Bumm, who likewise is now occupied at the University of Berlin. The three other daughters became wives of officers. Leube provided me valuable service in cases of illness and saved from death my oldest son Hermann, who suffered a severe infection of the bowel in early childhood.

Mrs. Leube had the good intention already in Erlangen to obtain a wife for me. She believed she had found the suitable girl in Miss Agnes Gerlach of Erlangen. However, my indifference in matters of love and the volley of scientific problems were not conducive to her plans. Finally, my illness and the fear of a relapse became a second roadblock. But women do not give up their pet ideas so easily, and so she arranged for repeated visits to Würzburg for the attractive designated miss. She was most strongly supported [in this endeavor] by Mrs. Dr. Knorr, who likewise had become friends with Miss Gerlach. During one of these visits, it actually came to an engagement between the miss and myself. It was on December 1, 1887, where I was 35 years and my bride 26 years old. The marriage took place in Erlangen on Saturday, February 22, 1888[2] shortly before Carnival, so that I had 4 days vacation to accustom myself to my new condition.

We had to remain in Würzburg a few weeks, and then middle of March at the start of the Easter vacation we took a four-week trip to Italy. I do not want to report details because the consummation of marriage is too intimate a matter. I can only say that my dear wife was a creature distinguished by physical beauty, purity of soul, and sweetness of temper. Her parents had done everything for her and thereby, perhaps, spoiled her too much; for the duties of marriage and leading a large household sharply curtailed her enjoyment of life in Berlin, and a certain indifference to her own wellbeing resulted, which had a lamentable influence during her final illness and perhaps [hastened] her death. She died on November 12, 1895, in Berlin from meningitis resulting from a middle ear infection, probably because she refused the [life-]saving operation until it was too late. In Würzburg she gave me two sons.

The oldest Hermann Otto Laurenz was born on December 16, 1888, and was from the outset a strong, healthy child. This conformed also to his later development. He suffered only one dangerous illness at two years, a stomach- and bowel infection, which the family doctor foolishly treated at the outset with an antidiarrheal [agent] thereby causing a tenacious, dangerous bowel blockage. The child would surely have died from thirst had we not, on Leube's advice, finally supplied him anally with a large amount of water.

The birth of the second boy, which occurred on July 5, 1891, was premature, and consequently the child was weak. [He] recovered quite quickly, but when we moved to Berlin, the now entirely strong boy suffered from the bad milk of the large city. Consequently, in June 93 he contracted a persistent bowel infection and, in the autumn of the same year, scarlet fever followed by a very unpleasant bronchial infection. These circumstances possibly affected the child's nervous system unfavorably. Nevertheless, he developed into a tall, strong, and mentally agile youth who passed his school[work] with ease and moved on to the university at 18 years. Without allowing himself a recovery pause after the *Gymnasium*, he plunged ardently into the study of medicine. Everything seemed to bode well for a promising future, when in the summer of 1910 a precipitous impairment of the health [occurred.] He had stubbornly insisted to serve the prescribed half-year [in the military] as a doctor during the summer and unluckily [was posted] to an infantry regiment in Jena. During the forced military duty of the summer the unusually tall, not yet fully developed young man was overstressed. He developed heart trouble and was released from the military. The illness was in itself not bad, but it had a ruinous influence on the young man's temperament, for he saw the disease, as is not infrequent among young medical doctors, as a particularly darkening picture. He continued his studies and, in the spring 1912 in Heidelberg, passed his preliminary medical exam with distinction,

---

[2] Translator's note. February 22, 1888 was a Wednesday. Also, Ash Wednesday fell on February 15, 1888, so the date given in the text for Fischer's wedding is almost certainly incorrect. It would seem that February 11, 1888 is the correct date of the wedding.

studied then two semesters in Würzburg and, because of military things, returned in the summer 1913 to Berlin. But here he had a nervous breakdown. His strength to work was exhausted. He believed he was becoming sicker, allowed himself to be sent in the autumn to Nauheim, then later to Meran, and [he] fell into a deep melancholy. Neither I nor the attending physicians correctly recognized his condition, otherwise one might perhaps have averted the progressive mental illness. And so, it came to an obvious outbreak of mental illness in November in Meran. A stay of several months with Binswanger in Jena succeeded in [bringing about] a temporary improvement. He remained in Jena during the summer as a student. Although released from active duty, following an inquiry from the military authorities, he had committed himself to provide medical service in case of war. He was thus ordered in September 1914 to report [for service] as an under-doctor to the military hospital in Erfurt. He promised to do this and performed well at first, but after 5 months he had a dispute with his superior officer, and in connection with this he had a relapse of his illness. He went again to Binswanger, but he did not regain his strength to work. Sometimes it seemed that the illness was entirely gone, so quiet and reasonable as he knew to appear. When I took a stay of several weeks with him in St. Blasien, I was full of hope that the cure had taken hold. But suddenly a relapse came, worse than the previous one. We returned to Wannsee, and he soon went again to his friend Binswanger. Here his condition worsened, and he had to be taken to a closed institution. The awareness that now in all likelihood he was a lost individual led him in a state of deep depression to the decision to take his own life. He died on November 4, 1916, at the age of 25 years and rests in the small graveyard at Wannsee. A dear, good son, a talented and ambitious young man went with him.

My third son, Alfred Leonhard Joseph, was born on October 3, 1894, at Ambach on Lake Starnberger. At that time, I was already in Berlin, and my wife wanted to visit her parents in Ambach for a few weeks. Since she was in an expectant condition and had already once had misfortune, I believed that traveling was too risky. But considering the advanced age of her father, she would not be restrained. The consequence was that [on arrival in Ambach] she had to go straight to bed and lie there for 5 months until the birth of the boy. This one also came into the world very delicate. He later likewise became a powerful man, but he had very delicate skin and suffered from time to time with nervous headaches. In early youth during the summer at Wannsee, he suffered from a malaria-like condition twice in succession, until we moved our abode there to the hilltop. [Like his brothers] he was talented, diligent, and a good pupil. At 18 years he left the *Gymnasium* and wanted to become a chemist like the eldest brother. I advised him against this, however, because of the great sensitivity of his skin and his nerves. He therefore studied physics next but left it after two semesters and became a medical student. While he was studying in Heidelberg, the war broke out. He volunteered for service but was deferred. In January 15 he entered an artillery regiment in Berlin. After successful training he went into the field in September of the same year and served as a medic in the munitions convoy where his brother Hermann was lieutenant. Here he was soon promoted to non-commissioned officer. In July 1916, during his last leave in Heidelberg, he passed the preliminary medical exam with distinction. In August of the same year, he was sent with the munitions convoy to Rumania, and both brothers made the advance to Dobrudsha. Here he was promoted to field under-doctor. While the older brother took six weeks of training in Berlin to become a gas protection officer and then accompanied the advance of the German army into Rumania with von Falkenhayn's headquarters, Alfred [served] in several hospitals, finally, to his misfortune, in a military hospital for soldiers [with epidemic disease] in Bucharest, where the sanitary conditions, according to his own portrayal, were quite bad. Here he was infected with typhus, and after a 14-day illness he died on March 29, 1917. He was buried at the Memorial Cemetery in Bucharest, and his brother Hermann, who at the time was in Focsani, took a short leave to pay his final respects. He was likewise a very dear, reasonable, and gifted individual of distinguished character and very talented in interacting with people. Probably he would have become an excellent medical doctor, possibly also a successful researcher.

In the summer 1888 the elderly Robert Bunsen retired from his teaching post. The professorship was offered next to Victor Meyer, but after some hesitation he declined and wanted to remain in Göttingen. After that I received the call, and the relevant department head from the ministry of Baden came to Würzburg for the negotiation with me. The particulars were in general quite favorable, and although I very much liked Würzburg, Heidelberg held a certain attraction for me and even more for my wife. We therefore drove to Heidelberg in the spring 89 to learn about the details. At the hotel we soon met Excellency Bunsen, who had given up his service apartment and had for a time taken up residence at the guesthouse. The venerable old gentleman received us with great politeness and attempted to make the advantages of the Heidelberg position as clear as possible. After the first conversation I asked my wife what impression she had received of the great chemist. She responded laughingly: "First I would like to wash him and then kiss [him;] for he is an entirely dear man."

The next morning Bunsen showed us the laboratory on the Wredeplatz that he had built and used for so long. He was entirely in love with the old building, which of course bore the solemnity of a great tradition and intense scientific work. However, the resources were quite modest by comparison with the current [standard] at the time. Ventilation was still provided, as in the old alchemist kitchen, through a large chimney hood. When I asked if this were sufficient, the old gentleman declared: "We have the pure garden air here." On the contrary, the assistant, whom I turned to in confidence, indicated that the stench was usually unbearable.

At the end, when we also wanted to inspect the apartment, Bunsen led us through the entry door, then pulled out a gigantic ring of keys and tried each individual key to see if it fit. The attempt was in vain, and the result became no better on repeated attempts, so that finally I had to ask the old gentleman to cease his efforts. I learned afterwards that the operation with the keys was an act. Bunsen simply did not want to show us the apartment, because he feared that my wife would not like it, just as Mrs. Meyer had not, and this would constitute a reason for declining the position. Of course, he was mistaken in this. It was not the apartment, but rather the measures taken to expand the institute, which in my view amounted to mere patchwork, that were decisive for me to remain in Würzburg, after a new building for the institute there was in prospect.

In the meantime, Meyer had had a fundamental change of heart. He terribly regretted not going to Heidelberg and immediately visited me in Würzburg in order to clear this up. We were soon of one mind, for I preferred to be able to decline without complaining to Bunsen and the Heidelberg faculty about the lacking estimation of the professorship there. The matter was settled by telegram with the ministry in Karlsruhe, and Meyer then, as is known, went to Heidelberg. The old Bunsen laboratory remained in service but was supplemented with a new building for the organic division. I later looked at the expanded laboratory and became convinced that my original opinion was correct. For the same money one might have built a larger and more serviceable building 10 min from the city.

The university administration at Würzburg and the culture ministry in Munich immediately gave their thanks for my remaining; for they proposed the new building for the chemical institute, for which the city placed a splendid building site on the Pleichering at [our] disposal in exchange for the transfer of the old building on Max Street. Also, there was the prospect for me of a raise of 1000 marks, without my having made such a request. But both items needed approval from the Bavarian legislature. They were abruptly denied by the ultramontane majority. The salary increase was even denied a second time when the Chamber of Imperial Advisors reinstated the small sum into the budget. The reason for this poor treatment became known to me only several years later. As a consolation a Bavarian order [of merit] was bestowed on me and soon thereafter the salary increase was also granted, after the means had become available through a case of death. The request for the new building, in the amount of 650,000 marks of course had to be postponed 2 years, that is, until the next budget cycle. The affair then took a very amusing and, for Bavarian circumstances, so characteristic course, that I want to convey the details here.

In order to examine the necessity, emphasized by the government, of a new building for the chemical institute at the location and [building] site, a commission of the legislature appeared in Würzburg. It consisted of the ministerial official Dr. Bumm, the representative of the ultramontane majority Dr. Daller, and the representative of the liberal minority Dr. Schauss . [Schauss] was decent enough to look me up prior to the official visit, since he knew me personally from the Munich time. He opened the interview with the question: "How does it happen that you, as a good Catholic man, have your children baptized as Protestant?" To my clarification, that I had always been Protestant, he replied: "Good God, then a severe injustice occurred for you two years ago," whereby he referred to the poor treatment on the part of the Bavarian legislature. When I then added that my wife was Catholic, he declared laughingly: "Well, that is so much the worse, you therefore earned the punishment at that time fairly."

Soon thereafter we had the honor to receive the commission. In order that this occur in the most worthy manner, after agreement with the assistants and students, preparations were made, not with flowers or white-clothed virgins, but rather in a far more effective manner with the strongest-smelling materials of chemistry. Bromine, hydrogen sulfide, ammonia, mercaptan, skatole, isonitrile, cacodyl had served to fill the various rooms of the institute with an infernal atmosphere in order to demonstrate convincingly by nose to the members of the commission the insufficient space and poor ventilation. I can still see the astonished faces of the gentlemen, who bravely worked through the smell, and as we finally arrived in the cellar, the entire party took a deep breath and declared that the air here was much the best.

My goal, naturally, was to win over the gentleman Dr. Daller to the new building, and I put forth every bit of eloquence I could muster. As a clever man, he felt obligated to consider less expensive possibilities, for example, an addition to the old institute. As I pondered this thought, he posed the surprising, but certainly not unjustified question: "Who will guarantee us that such a donkey show does not happen again?" I was self-assured enough to deny resolutely personal [involvement in] the donkey show and had the impression that Dr. Daller as well as the other gentlemen were satisfied with my exposition. In fact, several weeks later the sum for the new building was approved by the legislature without any cutback, and I was fairly proud of my efforts in this apparent success.

However, when I visited Adolf von Baeyer in Munich half-a-year later and related the course of the negotiations over the new building of the Würzburg institute, I experienced a certain feeling of disappointment. [It seemed that] Dr. Daller's change of heart had not been caused by my eloquence and other meetings, but rather through the intervention of one of my old pupils from the Munich time, Dr. Brandl, whom I had directed, as previously mentioned, in a small project on the determination of fluorine in silicate. This gentleman had become close friends with Dr. Daller and made serious reproaches because of the poor treatment I had received on the part of the Bavarian legislature two years earlier. That was the reason why Dr. Daller now stepped in and endorsed approval for the undiminished building sum.

For the initial request to the legislature the university architect von Horstig and I had drawn up a provisional building plan and now suddenly received from the government the surprising mandate to carry out this plan with the appropriated sum. I flatly declared to the university administration that the plan was thoroughly inchoate and must be replaced by a new project thought through from every direction. But the university architect had no interest in doing this, because in the meantime the assignment had fallen to him to build a new building for the entire university. For the institute, therefore, it was necessary to find a special architect. This assignment again fell to me, and through diligent inquiries of experts in the field, I succeeded in locating a splendid young architect in the Bavarian civil service. My suggestion to entrust this gentleman under favorable circumstances with the institute building was met with great cooperation from the culture minister in Munich. But the building administration felt that their rights were infringed by my course of action and flatly denied my request. Consequently, I turned, with the agreement of the university administration in Würzburg, to the outstanding architect Professor Hase in Hannover. I had previously exchanged letters with him concerning his son, who was studying chemistry in Würzburg. He recommended to us a splendid young architect from the circle of his pupils, and a contract with this gentleman came about. But when this was presented for approval to the authorities in Munich, there arose a great excitement over the choice of a Prussian architect and over the high-handed course of action by Professor Fischer. The contract was annulled, and the Munich building authority sent, as they themselves could spare no suitable man, a young Swabian architect to Würzburg, who later proved to be incompetent. The entire dispute with the building authority in Munich did, however, have the benefit that after their serious remonstrances, now the splendid university architect von Horstig declared himself ready to undertake the new building of the chemical institute. Together with him I drew up a new plan, which was, according to my feeling, thoroughly thought through and, in the proportions of the space and the technical furnishings, corresponded to the needs of the University of Würzburg. With slight modifications, it was in fact carried out, to be sure, only after my departure from Würzburg. I believe that even today the Würzburg institute still belongs among the best-constructed laboratories in Germany.

Out of old habit, I regularly used the vacations in Würzburg for shorter or longer travel. At Easter I mostly went to the South. The trip to Corsica 1886, where the skatole smell accompanied me, is already mentioned earlier. After 14-day stay in Ajaccio, I received a visit from W. Königs and R. von Pechmann who came from Nizza and landed on the island somewhat shaken after a stormy night. Several days later, we drove together in a private wagon from Ajaccio diagonally through the island via Corte to Bastia. It is a stretch of perhaps 120 km, and the drive lasted two days. It was a respectable accomplishment for both small Corsican horses, for the beautiful road leads over a considerable [mountain] pass height of over 1000 m. There were no railroad connections on the island at that time.

Because of the barrenness of the land, we arrived in Bastia somewhat famished. The journey continued on to Livorno the same night. Here we were compensated for the small deprivations of the otherwise enjoyable country drive with an opulent meal that Pechmann together with the *chef de cuisine* of the hotel jointly prepared. Pechmann was not only a good chemist, but also well-schooled in the art of cooking. Later I took several more trips together with this meritorious colleague. Königs dedicated an excellent necrologue to him. Here I can only confirm that he was a very pleasant travel companion who gladly took on the role of travel marshal and whose directions one could calmly trust, if one were not on a budget.

The following year I spent the Easter vacation at Bordighera on the *Riviera di Ponente* in the company of Baeyer, and later again with Baeyer and once also with Victor Meyer. In our circle there also was the poet Ludwig Fulda, who entertained us mightily at table with all sorts of jokes and anecdotes. The days served regularly for strolls and outings in the splendid surroundings, and the evenings we spent just as regularly gambling in the beer bar of the hotel, where the largest wager, to be sure, was set at 2 *soldi*. Baeyer was the expert, as he had once made a stay of several weeks in Monaco and had mingled with the croupiers of the casino. He was therefore the banker, and it gave him great pleasure as such on many an evening to rake in winnings of 4 to 5 francs. I myself have always been an enemy of gambling, because it was rebuked so often by my father as one of the worst vices. However, I do remember the evenings of gambling in Bordighera with pleasure, where, to be sure, the passions were not too greatly aroused.

For many people, of course, such a fling with gambling is a dangerous thing. Our friend Pechmann suffered heavily from it, as the greater part of his substance was sacrificed to the bank at Monte Carlo. Also, in later years, when he was in the vicinity of a casino, he needed close supervision from his friends lest the travel fund be placed in danger. I also encountered such friends of gambling among my later colleagues in Berlin, who otherwise were very clever and reasonable people. For the passions are fairly independent from reason.

Naturally, there was also no lack of scientific conversations during our stay together on the Riviera, particularly during the strolls, and there was scarcely any important problem of chemistry that we did not deal with. In particular, the memory of a question of stereochemistry remains with me. In the previous winter 1890/91 I had been occupied with the task of clarifying the configuration of sugars, without entirely achieving the goal. The thought came to me in Bordighera to make the decision of the configuration of the pentoses through their correlations to the trioxyglutaric acids. Unfortunately, because of lack of a model, I could not determine how many such acids would be possible according to the theory, and I therefore presented the question to Baeyer. He seized such things with great warmth and constructed carbon atom models from toothpicks and little balls of bread. But after much testing he also gave up, apparently because it became too difficult for him. Only later in Würzburg, after lengthy consideration with good models, was I successful in finding the definitive solution.

Later I went several times on Easter vacation with Baeyer to Territet, near Montreaux, on Lake Geneva, where we once met also with Carl Graebe, who was professor in Geneva at the time. The stay in the sunshine on Lake Geneva was unusually refreshing, and the marvelous surroundings gave ample opportunity for beautiful strolls and larger outings. Graebe was a stimulating character in our circle, and when his clear, loud laughter rang out at the long table of the Grand Hotel, the attention of the entire company was drawn. I had so taken a liking to him, with his cheerful essence, his agreeable colloquialisms, the clever judgment in scientific and also purely humanitarian things, that I gladly strived for his friendship. In later years I repeatedly met with him in the Black Forest during the autumn vacation. At age 65 years he left Geneva, where he had worked as professor of chemistry for more than a quarter century and moved back to his home city Frankfurt-on-the-Main. He is now elderly but remains fresh in body and spirit. His great scientific works belong to history. But now he is also active as a writer, and in the previous autumn in Baden-Baden, I learned to my great pleasure that he has written a history of organic chemistry, which will surely complement the similar superb book by Ed. v Hjelt in propitious manner. I hope that the printing of this book will not be hindered by the war.

In connection with the stay in Territet, Baeyer and I took part in the 1892 International Congress of Chemists for Reform of the Nomenclature in Geneva. This [congress] distinguished itself favorably from the clamorous and confusing international scientific or technical meetings which took place in the last 20 years and which I avoided whenever possible. The [meeting] in Geneva consisted of perhaps 60 persons, who were all accommodated at the same hotel and who lived together for several days like a large family. Various ladies, among them also my wife, took part in the modest, but cozy social events.

The head of the Congress was Charles Friedel from Paris, a native of Alsace and a very genial man; I already knew him from a previous visit to Paris. He greeted me affably, as was his manner. We conversed at length, because I was able to relate to him much concerning his home city Strassburg. He was nearly moved to tears complaining about the separation of his home from France.

Most of the other European countries were represented. Adolf Baeyer played a leading role in the proceedings. He had likewise occupied himself many times already with the nomenclature question and succeeded throughout in having his suggestions accepted. I have little cause to report the particulars, as the proceedings are portrayed in detail in the chemical journals.

Much has remained of the consequence of the chemical congress and indeed become an enduring legacy of the chemical language. But the consequent implementation of a rational nomenclature according to chemical structure proved to be impossible, because in the end it led to names that were unusable due to their length. Also, that which one had chiefly envisaged, the registration of the carbon compounds with the help of such names, was, as we know, separated in the meantime through the practically simpler registration according to empirical formula, as it was first employed by M. M. Richter. However, here also the number of isomers listed under the same formula grows with frightening speed, and one must already think about finding a second means of registration, along with the empirical formula, to ease the search for the individual material.

The days of the Geneva congress will nevertheless remain [among] the best memories for all participants. Through the harmonious flow and the comfortable interaction, it showed a worthy likeness of the general interests, which should unite the representatives of the science in all countries. After the lamentable experiences of the world war, I harken back gladly in my memory to these better times and hope that with the return to peace, the reason and feeling of solidarity among scholars and in particular among natural researchers will also return.

During the Würzburg period I mostly went during the autumn vacation to a resort on the sea in Belgium or Holland and later to Norderney. On one such spa trip to Scheveningen 1889 my wife also took part, and we spent several enjoyable weeks there with Arthur Dilthey and his wife and afterwards visited the larger Dutch cities. On this occasion I also showed my wife Euskirchen and surroundings and presented her to our numerous family in the Lower Rhine. Usually, though, she went with the children to Ambach on Lake Starnberger during the autumn vacation to her parents.

Since my illness in Erlangen, I no longer came to the hunt in Euskirchen, which earlier I had found so refreshing, because I did not want to expose myself to a new illness. Instead, in September I repeatedly attended the meetings of the natural researchers, during the Würzburg time, those in Berlin 1886 and Heidelberg 1889.

The latter was weighted toward physics and chemistry; for Heinrich Hertz and Victor Meyer held both of the main sessions, the first on his great discovery of the electrical waves and the other on the general problems of chemistry. Also H. von Helmholtz and Werner von Siemens came to know Edison, who had just invented the phonograph and demonstrated this remarkable instrument through the help of an assistant. Further, I heard the physicist Bolzmann [*sic*] from Vienna speak for the first time. He had the peculiar habit of beginning every sentence in the highest voice register and ending in the deep baritone. This was so stressful that already after a few minutes the sweat was running over his face, and we listeners had to suppress our amusement despite the great respect for the speaker. Bunsen was not present. Probably he still had enough from the exertions of the 500-year anniversary of the Heidelberg University, which had taken place several years before in August 1886, in which I also took part.

At that time, I made my first visit, accompanied by Baeyer, to Bunsen, who received us in a most friendly manner and with mountains of cigars, but because of his deafness he was rather reticent in conversation. We met Sir Henry Roscoe there, who had hurried over from England in order to ease [the hosting duties] of his old teacher and friend Bunsen during the days of celebration. Roscoe was a very amiable man, who knew to adapt to us young colleagues. He entertained us by relating anecdotes from his student time at Heidelberg or from his experiences in England. I have seen him again a few times in England, and he has also sent me several very friendly letters. Naturally, at that time there was no lack of the usual celebrations at the old castle. There was indeed an enormous drinking bout, at which the archduke himself presided, and similar academic events.

The natural researcher meeting in Berlin from the year 1886, the first in the new imperial capital, was unusually well attended. It was of a different character than Heidelberg but was also very interesting for us chemists. The events in our section were rich in scientific content, and local members of the chemical society took pains to enhance the stays in Berlin of their out-of-town colleagues through cozy interaction and organizing merry celebrations, for example a beer evening of the thirsty chemical society under the guidance of C. Scheibler. Here I met A. W. von Hofmann and admired his proficiency in business and characteristic matters.

On this occasion, the Chemical Society had organized an exhibit of scientific preparations, to which I contributed, among other things, the newly prepared synthetic indole. There was also a large amount of entirely pure skatole, and the discoverer of this material, Professor Brieger, was very pleased that my preparation also had the evil smell he had described; for A. von Baeyer had communicated sometime earlier that pure skatole does not smell unpleasant. He had apparently fallen victim to the not infrequent deception that can arise from the various effect of smelly materials in concentrated or dilute form on the olfactory system.

At the beginning of the year 1890 I was able to communicate the synthesis of mannose and levulose [fructose] in the journals of the Chemical Society. The result was an invitation from the chair to give a summary talk on carbohydrates in Berlin. This took place on July 23, 1890, and I was able to illustrate with experiments, supported by my co-worker J. Tafel, the important phases of the investigation. It was the first time that I spoke at the Chemical Society, and I earned as thanks a few very friendly words of recognition from the chair, Mr. A. W. von Hofmann. Afterwards, as was and as remains the custom, an evening meal in simple form to honor the speaker took place.

When I returned to Würzburg, my wife, who was curious as to the course of the talk, asked me about it. I allowed myself a little tease and told her the people had called "Au" during the main passages of the talk. There followed great disappointment and remonstrations against the impolite Prussians, which she as a Bavarian felt fully justified [in making;] for she has also called me a Prussian in anger from time to time. Of course, I then explained the joke, but the good mood had been spoiled.

The first syntheses of natural sugars also brought me the first public recognition from abroad; for soon thereafter I received the Davy medal from the Chemical Society of London and was elected as a corresponding member to the scientific society at Upsala.

Although Würzburg did not lie just on the great army road, nevertheless I was allocated many a dear visit from colleagues there. Eduard Hjelt came one day as a new student to the university, later Victor Meyer, then Otto N. Witt, H. W. Perkin, Jr., and many others. I was most surprised by Ernst Haeckel from Jena, who appeared early one morning with travel bag [in hand] and was in a great hurry to enquire after Ludwig Knorr. From the brief, but very lively conversation, one expression of Haeckel's has remained in memory: "When you chemists synthesize egg white, that will scratch an itch." Knorr in fact received a call to Jena a few months later and moved there in the autumn 1889. It is now nearly 30 years that he has been a respected member of the Thüringian institution.

The most curious visit I received [was] from America. One day a professor of physiology appeared, who had received money from a wealthy man to establish a university in Worcester. He had the romantic idea to load an entire ship with European professors, assistants, instruments, preparations, and similar things and then outfit his university with the apparatus. He opened the conversation with me with the question: "Do you want to go to America with me as professor?" I was so surprised at this that I held the entire [thing] to be a joke until he developed his detailed plan. He was, by the way, a refined and widely traveled man, who had much of interest to relate.

Soon after that there appeared an American lady who presented herself as Miss Helene Abott [sic] and a colleague. For her special protection she had brought along a second female individual, who on closer inspection turned out to be a Negress. She declared that she wanted to study science in Würzburg and was astonished that women were not yet permitted to [attend] lectures. I showed her the laboratory and presented her to the young gentlemen Knorr, Wislicenus, Tafel. She made entirely knowledgeable remarks and showed that she had no mean theoretical understanding. After her departure a war council was held, [to decide] whether we should procure entry to the laboratory [for her] from the University Senate. A few were enthusiastically in favor, but the prudent individuals could not suppress the fear that she could easily bring about confusion in our circle, which up to then had been so harmonious. Corresponding to the majority decision, I wrote her [a rejection letter] and received from her a polite, to be sure, but rather energetic answer, wherein she reproached the backwardness of Germany in the education of women. She later became the wife of Arthur Michael, but, as far as I have learned, they went their separate ways after some time.

In the spring 1892 I had to lie in bed for several days due to a bout of influenza. My wife read aloud to me, right from the journals of the Chemical Society, which had recently appeared, the necrologue of Peter Griess,[3] about whom I wrote the scientific part myself.

But much more interesting was the personal part written by A. W. von Hofmann and spiced up with delightful humor. We were just laughing heartily over it, when a telegram from Tiemann arrived, bringing the news of Hofmann's sudden death. Because of my illness, I could not go to the funeral, which I regretted all the more, since I belonged to the executive committee of the Chemical Society and had been received by Hofmann in so friendly [a manner] during my most recent visit to Berlin.

The condition of my health at that time was not satisfactory. As I indicated earlier, I had fallen victim to a chronic poisoning with phenylhydrazine. While many of my co-workers and also several servants were very sensitive to the base and reacted with nervous trouble or with severe swelling of hand and arm, I seemed to be very resistant against the poison. Until 1891, the damaging effects were limited to eczema of the fingers and palms. The chronic poisoning, which appeared in the autumn 1891, manifested so much worse as a very troublesome disorder of the gut, namely with nightly colic and diarrhea. The illness reached its apex in the winter 1891/92 and mocked all normal treatment by physicians. Only the application of Priessnitz wraps brought me relief and the sleep I had been deprived of for so long. The poisoning occurred in part through vapors, but, as I was later able to determine, much more through the skin, that is, from the hands. I have suffered from this for many years and finally developed an idiosyncrasy against phenylhydrazine and similar materials. It was the second injury that came from my profession, and I would probably have succumbed to it, had the reason not become known and had I not from that point on avoided contact as much as possible with the harmful base. The poisoning naturally also had quite a bad effect on my nervous system. Staying in the service apartment of the Würzburg institute, which was continuously filled with fumes from the laboratory and was also unusually hot, became so unpleasant in the summer 1892, that I moved my household to a rented house in the country with large garden.

Here, on a beautiful June day, there suddenly appeared Privy Councillor Friedrich Althoff from the culture ministry in Berlin. In apparently entirely naïve style, he related to me that he wished to use this coincidental stay in Würzburg to renew our acquaintance from the natural researcher meeting in Berlin. He was very pleased to give his opinion on the simple south German lifestyle [and] on the modesty of

---

[3] Translator's note: The chemist Peter Griess died in 1888.

professors in these parts, then [he] came to the circumstances in Berlin, the chemical institute there, and the intention of the culture ministry to do as much as possible for the cultivation of chemistry in Prussia. [He went on to say that] it would interest him also to hear my views on this, whereupon I declared frankly that the institute built by Hofmann no longer suffice in any way for the present. Only now at the end did he pose the question whether I would not want to attend myself to the necessary new construction as successor to Hofmann. The faculty had proposed me along with Kekulé and Baeyer but expressed the wish, due to the advanced age of both of the others, that I in fact be offered the post. By no means was I overjoyed with the offer, which in itself was really quite glorious and also made in such obliging fashion; for it would then be necessary for me to choose between Würzburg, where I felt so happy, and Berlin, which dawned before me.

My decision would have been reached quickly and made in favor of Würzburg, had I stood alone and followed only my feeling. But my wife was more ambitious, and I had to give Althoff at least the promise to come to Berlin to become acquainted with the circumstances of the place and the position. This came to pass 8 days later. At the ministry in Berlin, they were most obliging in every respect. The minister Excellence Bosse, due to lack of time, received me on Sunday at 8 o'clock, in order to assure me that everything would be done to fulfil my stipulations, particularly also the new construction of the institute. Also, the Berlin colleagues strongly encouraged me to accept the position. Having still not decided by any means, I drove to Munich, where I had been invited by the minister there. I was astonished by the clumsy manner in which he tried to convince me to decline the Berlin offer. First, I had to wait 1½ days before he even received me. Then he maintained that through the approval of the new building in Würzburg, I was obligated to remain there. I answered him that the building had not been approved for me personally, but even if it had, one could [cancel] the project, since construction had not yet begun at all. In short, I returned to Würzburg somewhat disgruntled. In the meantime, my old father, who had heard of the Berlin offer, had arrived [in Würzburg] and immediately sprang to the fore to encourage me not to decline thoughtlessly such a good business. This position in Berlin would not be offered to me a second time, [according to him.] Furthermore, I could always change later if I didn't like it there.

Also, I could not conceal from myself the advantages of Berlin. The lively scientific life of the imperial capital and the foreseeable means for the possibility of assembling a larger circle of pupils around myself in fact held great attraction for a man my age (I was not yet 40 years old.) And so, after 8 days of vacillation, I came to the decision to set aside my personal inclination and accept the position. Then I drove for a second time to Berlin, accompanied by my wife, to rent an apartment, as the service apartment was still occupied by Mrs. von Hofmann until May 1893. Also, [there was] a series of small structural changes to arrange, which would be seen to during the autumn vacation.

After I was firmly committed, several colleagues spoke more openly about the conditions in Berlin. My old teacher Kundt surprised me with the remark: "Well, Fischer, you'll soon be surprised by the stack of work they load on a professor here." Somewhat terrified, I asked him the question why he had not said this to me 14 days earlier, as I had relied on his old friendship for clarification of the conditions in Berlin. He responded laughingly: "Well, then you would not have come."

At this last stay I also enjoyed the first test of society in the Berlin circle of scholars. Mrs. von Helmholtz had received word of our impending visit to Berlin and invited us by telegram to company in the evening. This took place in the splendid service apartment of the Physical-Technical Imperial Institute in Charlottenburg, and there we met an interesting circle of people, including the old Werner von Siemens, with whom I conversed at length about technical-electrochemical problems, and who honored my wife afterward through special charm. I had to give my first after-dinner speech in Berlin here in response to a few words of greeting that Helmholtz made to my wife and myself. Moving outward from Kepler's laws of planetary motion and their influence on the development of physics, I was able to develop an astronomical portrait of the company, at the center of which came Mrs. von Helmholtz as the sun. Apparently, I won her favor thereby; for she was later always obliging to me in a very friendly manner.

On the journey away from Berlin I was in a very melancholy mood, caused by the poor condition of the chemical institute and by the surprising information about the obligations that Hofmann's successor expected. If I had not been ashamed to break my word, I would have sent a telegram on this return journey to the culture minister in Berlin to withdraw my acceptance. But it was now too late for that, particularly also because I had the feeling that at the ministry in Munich, they were peeved at my acceptance of the Berlin position.

My father in the meantime had left Würzburg, entirely satisfied by the success of his persuasion. For himself a change of domicile was a mere trifle. He had recently decided to leave Euskirchen after a 56-year stay and move to Strassburg-in-Alsace.

For me now began an uncomfortable time, the preparations for the move to Berlin. A professor of chemistry moves not just with his scholarship and the books, but also with preparations, apparatus, and assistants. [The assistants who] followed me included Dr. Oscar Piloty, the previously-mentioned Dane Dr. Fogh and Dr. Lorenz Ach. Wislicenus had in the meantime become associate professor and as such was bound to Würzburg. For Julius Tafel, whom I would gladly have taken along, a suitable position in Berlin could not be found. Against that, I was happy that the servant J. Wetzel, who later became well-known through several useful [pieces of] glass apparatus, was able to transplant to Berlin as preparator. I had nothing to do with moving the household because my wife assumed this as her right and her duty.

The departure from the dear city Würzburg, the colleagues and students was cordial but brief. The two children came to the grandparents at Lake Starnberger, and I moved with my wife and a cook to a small, absolutely delightful wooden house, 10 min from Berchtesgaden in the Untersberg, set in splendid surroundings. Here we lived, as on our honeymoon, entirely for ourselves, and we spent 6 weeks in magnificent weather. It was the right preparation for the coming Berlin period. It is true that the German Chemical Society, for the celebration of their 25th anniversary in November of the same year, had entrusted me with the remembrance speech for A. W. von Hofmann. But it soon came to my attention that F. Tiemann, as friend and pupil of the deceased, would be much more suited to this task, and he was also glad to assume it. With agreement of the head of the Society, we exchanged, and the release from the speech for me was, to be honest, an agreeable relief.

Fig. 8.1 In the laboratory

Chapter 8
Berlin

In the middle of September my wife went back to Würzburg to arrange the move to the Berlin apartment, which we had rented for a year in the house of a Dr. von Dechend, coworker of Tiemann according to the reports, in the Queen Augusta Street near the *Tiergarten*. At the end of September, I also left Berchtesgaden and met my wife in Berlin. She was entirely exhausted and in tears, brought about by the disagreeable interactions with the Berlin movers. As a South German she was not prepared for these rude people, and only with the aid of Dr. Piloty, who impressed the Berlin movers with his massive bodily size and his energy, was she able to get through with her wishes and directions. Then there was the melancholy mood that weighed upon the imperial capital at that time because of the great cholera epidemic in Hamburg, which necessitated a strict vigilance of personal interaction. The skillful accomplishment of this sanitary measure, which protected Berlin from the epidemic despite some 100 cases of cholera [brought in from outside the city] impressed me greatly in retrospect. But for all people who immigrated to Berlin at that time from the simple south German circumstances it was an uncomfortable time. My wife was dismayed for a second time when soon thereafter she fetched the children from the grandparents on Lake Starnberger; for the nanny, whom she brought along, could not stand the train trip, fell ill along the way and was placed in supervision from Leipzig on.

Thanks to the help of our friends, the settling-in of the household and the family into the Berlin circumstances proceeded quicker and better than we thought. Also, the small renovations in the laboratory, which in particular were designed for better ventilation, were finished on time. And in the last weeks of October the institute could again be opened for regular business.

Tiemann, who up until then had had his hands on the leadership of the [teaching] laboratory, retreated from this [duty] but remained at the institute as private scholar for the advancement of his successful scientific studies and also held a small special lecture [series.]

I was easily able to come to an agreement about the division of work with Gabriel and the other assistants inherited from Hofmann, Dr. Pulvermacher and Dr. Richter.

© The Author(s), under exclusive license to Springer Nature Switzerland AG 2022
D. M. Behrman and E. J. Behrman, *Emil Fischer's "From My Life"*,
Springer Biographies, https://doi.org/10.1007/978-3-031-05156-2_8

The assistants I brought along, Piloty and Fogh, took over the instruction in the main [lecture] hall, which was assigned to the analytical division. Gabriel remained in the organic division, and I myself assumed oversight of the entire operation, with exception of the two large experimental lectures. Dr. Lorenz Ach stood at my side for my own investigations as Private Assistant, a post he had already held in Würzburg. The only Würzburg student to move to Berlin with me was the Englishman Crossley, who later became professor at the pharmaceutical institute in London.

I began the winter lecture [course] on inorganic chemistry with a brief obituary speech to my great predecessor. The format, similar to what was used in Würzburg, was entirely different, though, because this scheme was more easily understandable and better adapted for an audience consisting mostly of medical students, apothecaries, and teachers. It gave me particular pleasure, though, to have the assistance of Dr. C. Harries, who had already been Lecture Assistant with Hofmann; for I was able to incorporate into the presentation the experiments developed by Hofmann along with ones I already used. And thus, the extensive book of lecture experiments arose which, with useful supplements, has been used since then in Berlin and has served as a model for many younger colleagues.

Scarcely was the institute in full operation when a small revolt of the servants broke out. They suddenly declared [they] could not and would not perform the work. Apparently, they had misunderstood the somewhat too friendly manners to which I was accustomed in southern Germany and now held a test of power towards me as appropriate. In addition, they were somewhat spoiled by the previous regime regarding accomplishment of work. I now had to change the tone. I immediately fired one servant, who was terminable; another, who unfortunately was permanently employed, was retired by the minister; and I quickly procured other help. In Berlin this was the right method, and I never again had serious disagreements with the servants, for whom, by the way, I also strove to the best of my ability. On the contrary, I have to say that with proper treatment they were more useful and more capable than those in southern Germany.

The chemical laboratory [building] built under the direction of A. W. Hofmann on George Street, which went by the name of the "1st Chemical Institute of the University," in contrast to the "2nd Chemical Institute" on the Bunsen Street, planned by Rammelsberg and later used by Landolt, was valued as a tourist attraction for its architecture and façade, but for chemical purposes it was quite impractical. Everywhere there was lack of air and light, and a large part of the space consisted of dark and unusable corridors. Only the two main work rooms on the first floor facing the George Street and the spacious private laboratory could be viewed as normal work rooms. Against that, the lecture hall was so dark that, except for midday from 11 to 12, it was necessary to use artificial lighting. The ventilation was also entirely insufficient. My first concern, therefore, was, just as at the Würzburg institute, the installation of an entire series of fume hoods in simple form, as previously described. In order to ensure adequate draft, a special air supply was installed, through a hole knocked in the wall and a wooden duct [leading] from inside the room upwards (Fig. 8.1).

Even the heating was in a sorry condition; for the existing peat ovens functioned so poorly that a fraction of the students had procured private heaters, which naturally became expensive for the institute through use of gas. They had to be replaced with new coal ovens.

The current budget, which previously amounted to about 15,000 marks, had been increased to 20,000 marks as the result of my request. Also, two new assistants were approved, but, as I later learned to my sorrow, [the positions were] snatched from the chemical institute at Göttingen.

For restoration of the inventory, I had likewise at my appointment stipulated a modest sum, 15,000 marks, if I am not mistaken. But when I wanted to make use of this [money,] there was a mighty clash with privy councilor Althoff at the ministry of culture. In retrospect, this sum seemed to him to be too large, and he even went so far as to propose that I should make these procurements out of my own pocket. It came down to a discussion in which Mr. Althoff was instructed that I was not in the least willing to allow myself to be mistreated and to renounce any stipulation of my appointment. We later had several more differences of opinion, but from that point on our conversation always proceeded in moderate form. Gradually he also became accustomed to offering me advice in chemical matters. He also had the irksome habit of keeping visitors waiting for hours in a quite uncomfortable waiting room. When this happened to me for the second time, I left. On his later inquiry, why I did that, I declared to him entirely frankly, that he might well make me wait if I wanted something from him; but if he wanted my help, he must receive me immediately, for my time be just as valuable and just as scarce as his. This he understood, and the longer I had dealings with him, the more I learned to appreciate him. He was a very clever man, full of ideas, who knew how to find a way forward for every worn-out jalopy. Then there was an unusual capacity for work and persistence in pursuit of his plans. Moreover, he allowed all important decisions to be determined only by pertinent considerations. He was not exactly concerned with outward appearances, and he offended several members of Prussian institutions of higher education with his unsophisticated manner. Nevertheless, I am of the opinion that his activity for the prosperity of Prussian institutions of higher education, in particular for their outfitting with institutes, libraries, [and] teacher training has been of great importance.

In spite of his openly rude manner, he was fundamentally a kindly man, who helped anywhere he could. This was evidenced in his solicitude for widows and orphans, for professors and servants weak with age, and the little-known fact that he donated a large part of his wealth for these purposes.

When my wife died at a young age, he took such pity on the three small children left behind, that for a full decade he visited them once-a-year and also prepared surprises for them more often by sending attractive books from the ministry. The boys were always overjoyed at this, because they believed that the culture minister himself was expressing recognition of them in this way for [their] good work in school. In concurrence with many colleagues, I held Althoff at that time to be the most important personality in the Prussian ministry of culture.

I also became well-acquainted with his wife, an Ingenohl by birth from Neuwied, whose mother, a von der Leyen by birth, was from the area around Flamersheim and was a youthful friend of my father. [This acquaintance developed] particularly during a stay in Meran, about which more later. She is a dear, clever woman and still lives, now very old, despite a severe heart condition, in Steglitz. She honored the memory of her husband through a very skillful biography with many very interesting original letters of the deceased.

The professorship of chemistry [at Berlin] had already been tied to several other offices during the tenure of my predecessor, and these were also passed along to me. First there was a chair at the so-called Pepinière, the current Kaiser Wilhelm academy for military medical education. This came with a small salary, and the students attend the regular lectures in experimental chemistry, just as do the other medical students. In addition, there is a scientific Senate at the academy, to which I was likewise elected, and to which I still belong. Here, under the chairmanship of Excellency von Schjerning, questions of various types pertaining to military medicine are discussed academically. The session closes with a merry evening meal in the splendid club room of the academy, which I always gladly attend.

I received the second medical office in the scientific delegation for the Medicinal World, which at that time was attached to the ministry of culture. The task of this delegation was manifold. [We] saw to the exam for the Prussian district medical officers and served as the highest expert authority for the Prussian civil- and court administration in all medical questions. Members included mostly clinicians, then public health specialists, and the chemists of the university. For all medical questions, therefore, an expert was present. If [a question] lay outside the narrow confines of medicine, it fell to the chemists. I had to offer an expert opinion in a variety of areas, sometimes on matters that I was unfamiliar with and for which I first had to seek the correct expert. [In one case] that remains in memory, I had to offer an expert opinion on the Queen Louise Spring, the so-called Health Fountain, which in earlier times lay well outside the city. In fine weather, citizens of Berlin would make a pilgrimage there to drink the famous water.

In the middle 90s, the water of this spring was brought to market by an apothecary as table water under the name "natural mineral water." However, since he had added carbonic acid, there followed a notification by the authorities because of fraud. The man was sentenced by the court to a small fine, but what was far worse, the police president issued a public warning about the table water, that it was not usual spring water. Due to this [warning,] the business was ruined in the briefest time. In his need, the owner turned to the Interior Ministry. And so, the matter came before the scientific delegation, and I was appointed correspondent. The exposition of the complainant was so clumsy that at first, I believed him to be in the wrong. However, I decided [to make] an inspection of the spring and secured the help of an official from the state geological office. After a long search, we found the once-so-famous spring, which had given an entire part of the city of Berlin its name, which, however, very few people knew anymore, in the cellar of a large apartment building. We recognized immediately that this was a genuine spring, rising strongly from a rather great depth, whose water was very good tasting and certainly hygienically harmless. Also, the saturation with carbonic acid was carried out in an expert and clean manner.

Consequently, my expert opinion on the spring was favorable. The police president had to withdraw his derogatory judgment, and the table water business was rescued.

From this example I have gathered how useful it is, when giving an expert opinion, not to rely on the dossier, but rather to become acquainted with the actual circumstances, wherever possible through an inspection.

Usually, though, my expert opinions for the delegation were quite boring and focused on trifles, for example, quarrels between apothecaries and drug store owners. I therefore preferred, after the course of my 5-year term, to leave the delegation. Landolt became my successor. Later, when Althoff became chair of the delegation, he called on me, semi-imperatively, once again as a member, but after several years I again left this post, and it was then transferred to the pharmacologist from the medical faculty.

The entry into the philosophical faculty [of Berlin], which took place at the beginning of November 1892, was carried out in simplest form, but with it came a pile of consequent duties; for the need for chemists for the numerous doctoral exams is particularly strong. At the oral exam for chemists Landolt and I were always evenly committed, but judging the written dissertations as the main referee was a [duty] that fell mostly to me, because most of the themes dealt with organic chemistry.

I was greatly surprised that conducting [faculty] business was so unwieldy. Since the faculty was undivided and at that time numbered 50 [full] members, now however, more than 60, one can easily imagine how detailed debates arise when so-called questions of principle are treated. Then there is the custom to undertake, not only all small matters, but also even voting on the outcome of every single doctoral exam by the entire body. This is linked to the establishment of the so-called Sedecim, that is, with the right of the 16 oldest members present to divide the doctoral fees among themselves. Entry into this circle is determined not by age, but rather by the number of relevant years the candidate has spent as full professor at any German university. No one can maintain that this patriarchal system is fair. I have always viewed this entire system as the main hindrance to all proposals for reform in conducting faculty business. Overall, clinging to statutes and traditions stands out so strongly in this body, that it often seems like an antiquated custom and at the same time just as laughable. For example, it was still the case on my entry to the [Berlin] faculty that the German states were deemed foreign; and as such, a resident could earn a doctoral degree in Berlin without having passed the *Abitur*. Only a year later was this nonsense done away with after a proposal by the physicist Kundt.

The size of the body naturally brings an enormous quantity of business, which takes much precious time to deal with. The philosophical dean is therefore a much-harried man. During the semester, as a rule, the faculty hold a meeting every week, which lasts two hours for the exams and approximately the same amount of time for faculty business. Anyone who takes part in the meeting of the Academy of the Sciences, then participates with the exams, has the pleasure to spend every Thursday from 4 to 10 sitting [in meetings.] Then there are the numerous exams for medical students, apothecaries, and teacher candidates. During the first 12 years of my stay in Berlin, I have sighed over nothing so much as the time and energy lost in this way. Later I succeeded in being released from this burdensome business.

Because of the size of the faculty, all important matters, and namely the business of appointing [new faculty] had to be referred to special commissions. Over the course of time, I belonged to many of these, and I was always pleased with the substantive approach to the proceedings, far removed from any type of clique [mentality]. Against that, I sometimes could not avoid the impression that appointments are not handled with the meticulous care that occurs at small and middling universities, where every individual to a man has an interest in the new colleague. This is clear in the custom, incorrect in my view, to appoint men primarily according to scientific reputation, but of an age where one no longer expects from them any great accomplishment either for the science or for the teaching. This holds true particularly for the natural researchers, who become exhausted earlier [in their careers] than the representatives of the humanities, and I have always felt obliged to raise misgivings against the appointment of old persons.

In general, the natural researchers do not play the role in the Berlin faculty that they could claim. The representatives of the humanities are more numerous and certainly more inclined to speech, perhaps also nimbler in form. Furthermore, they have more time and attend meetings more regularly and persistently, so they have the greater say. I have repeatedly had to raise objections against the infringement of the interests of the natural researchers. That has led occasionally to somewhat heated debates and raised some antipathy against me in the opposing party.

Chapter 9
The Academy of Sciences

Most members of the Academy are also professors at the university. The body first had personal connections with the technical college on establishing the technical division in the mathematical-scientific class in the year 1900. Then, as a rule, there are several men who do not practice a career in teaching, for the academy is entirely free in the choice of its members. In my time this last type of members was represented by the great electrical inventor Werner von Siemens, the botanist Pringsheim, the astronomers Auwers and Vogel, and later by the electrical engineer von Hefner-Alteneck as well as the railroad builder Zimmermann.

Even today the spirit of the Academy's founder, Leibniz, lives on. During the time I have been a member, now for over 25 years, I have always been pleased with the non-partisan and independent sense of this body. Every attempt by energetic men from the state government to influence the rules of its governance has been unanimously rejected. On the other hand, as soon as it came to a question of material wishes, [we] were dependent on the support of the state agencies, and I can say that during the period I am familiar with, the Academy has always found good will and vigorous support on the part of the supervisory culture ministry. Earlier it was Althoff who handled [our] affairs at the culture ministry, and in the last 10 years Fr. Schmidt, who today is himself the minister of culture.

The internal affairs of the Academy were handled, almost without exception, in a purely business-like manner and with refined calmness which not infrequently showed a touch of ennui. Heated debates arose only occasionally in the choice of new members. Only with the war has it become different. The war has, for the first time, also acted to confuse the mindset of many members, to be discussed later.

Although founded by Leibniz and reorganized in detail by Frederick the Great following the model of the Paris Academy, the Berlin Academy never lost its Prussian character. This is evident in the unusually painstaking and pedantic form of the affairs and still more in the stipulation that every member has to hold a scientific talk each year on a particular day determined by the so-called library card. It goes without saying that such a compulsion is inconsistent with the spirit of scientific research. It has also come to be felt by many a member, as known to me, to be quite

© The Author(s), under exclusive license to Springer Nature Switzerland AG 2022
D. M. Behrman and E. J. Behrman, *Emil Fischer's "From My Life"*,
Springer Biographies, https://doi.org/10.1007/978-3-031-05156-2_9

uncomfortable. However, repeated attempts to do away with this have foundered. Equally superfluous, in my opinion, is the provision that articles published in the Academy's [reports] may be published elsewhere by the author, within the statutory interval, only with the permission of the Academy. In any case, this [provision] is little heeded any more by natural researchers since the secretary's office handles violations of the statute very mildly. This is very sensible, for natural researchers these days cannot wait 1–2 years after publishing with the Academy before submitting results to a journal, since the reports of the Academy have too small a readership. It could well suffice [though] to establish priority in publishing. The less favorable stipulations of publication have had the consequence that I, like most other natural researchers in Berlin, have recorded only a small part of my investigations in the reports of the Academy, and that is probably the reason that reports of the Academy as an organ of publication have not attained the importance that reports from similar foreign academies have. In narrow circles we have often considered the question whether it would be expedient to allow a change in the editing of the session reports so that for all important investigations carried out at institutes led by members of the academy, short communications in the session reports could be made. But the unwillingness of the Academy to accept short, sometimes preliminary publications, that came to light particularly in the philosophical-historical class, has been too cumbersome for the alteration plans. Then there is the power of custom, not infrequently found in such bodies, which opposes anything new and can assume quite peculiar forms. One small example may illustrate this.

In the old building *Unter den Linden*, which is now replaced by a new building, the assembly rooms of the academy were cut to much smaller proportions. Lighting and heating were in approximately the same condition as 100 years earlier, and the consequence was that, particularly in the public sessions in the overflowing room, the air [quality] was dreadful. To many an academic [speaker], as well as the audience present, the mostly two-hour duration of the session became a small torture. [We] therefore decided to make a change. I had, at that time, acquired some knowledge through studies into the questions of heating and ventilation for the new building of the chemical institute, so I was commissioned, together with the physicist F. Kohlrausch, to plan such an installation also for the session room of the Academy. With the help of a talented engineer, this task was quickly completed. But when the plan was laid before the entire Academy for approval and thereby it turned out that the air in the room would be refreshed 2–3 times per hour, there arose an embarrassing concern about draft. Indeed, one respected member of the body declared that he would get cold feet. Despite the assurance that the air brought in from outside would be warmed and would entirely lose the character of wind by being dispersed through numerous small openings, our proposal could no longer be saved.

I had a second failure of similar type as member of a commission for the construction of the new Academy building *Unter den Linden*. Here our proposals for the heating and ventilation were carried out in reasonable manner, but I could not get through with the acoustics. Following my experiences with the new building for the chemical institute, I wished that the session room be outfitted with coffered wooden ceilings and also with as much wood paneling as possible, as these act as sound-boards. Although the members of the Academy this time were on my side, I did not succeed in overcoming the obstinacy of the architects. They maintained that they understood acoustics well, threw our advice to the wind, and [the result was that] the acoustics in the new building leave much to be desired.

In the summer 1893, soon after my admission to the Academy, I gave my first talk in the mathematical-scientific class on the synthesis of the alkyl glucoside, which became the point of departure for many other projects.

In the Leibniz session, which took place in July of the same year, I also held my entry speech, which is published in the session reports, and in which I not only indicated the direction of my own investigations, but also gave, in entirely brief form, an overview of the development of chemistry during the previous 50 years. Dubois-Reymond gave the response and greeted me as the second sugar chemist, the heir, as it were, to S. Marggraff in the Academy. From this I realized anew what a popular material sugar is. Sometime later, when I gave a talk on the stereochemistry of sugars, Helmholtz came to me to express his joy that chemistry could deal with such complicated questions of molecular structure. Naturally, the opinion of such a man for me and my science was glorious. At the same time, I noticed that he had only half understood the matter, because the facts on which the speculation was based were too unfamiliar to him. But with the perception of the genius, he had nevertheless recognized the great progress that the teachings of van't Hoff and Le Bel and the specific application to so complicated a structure as sugar had brought to chemistry.

Since then, it has been my task, as it is with all members of the Academy, to hold a talk once-a-year on a designated day, alternately for the mathematical-scientific class and for the entire Academy. I have always taken the trouble to treat general problems close to my area of work as much as possible in a popular form and also, as a rule, to catch the attention of the audience. The experimental work published by me at the same time in the session reports does not give an accurate picture, as it was just a small extract from the talk.

In earlier times, where the program for public sessions was expanded through scientific talks, I spoke, at the invitation of the Academy, as the first of the mathematical-scientific class in the Friedrich session January 1907 on the chemistry of proteins and their importance for biology.

The business of the body is handled, for the most part, according to the wishes of the entire [body,] by four secretaries; but where expert advice is needed, the individual members also step into action, either individually or as part of a special commission. Where chemical advice is needed, the membership has never neglected to ask me. One of my first pieces of advice unfortunately had an embarrassing consequence.

The Academy is regularly consulted for recommendations on the conferment of the Order *pour le merite*, civil class, on foreign scholars. When it came to selecting a chemist, I felt obligated to recommend first L. Pasteur and then Frankland and van't Hoff. Academy and government followed [my recommendation], and the Order was offered first to Pasteur through the German ambassador in Paris. [Pasteur] declined, as was his right. However, at the same time, he committed the indiscretion not to keep his decision secret. The matter went public and was exploited by the French press at a large jingoist demonstration. Several years later, when I visited Paris to tour the laboratories there and to look up H. Moissan, he immediately put to me the question, why the Berlin Academy no longer elect French members. I answered that the behavior of Pasteur was to blame and [also] the concern of Berlin scholars [to avoid] further scandals, which could only damage the reputation of the science. He acknowledged and emphasized that the great majority of French researchers also hold it as incorrect to allow nationalistic tendencies in science. Since then, an entire series of Frenchmen, among them Marcellin Berthelot, on recommendation of our Academy, have received and accepted the Order.

Another proposal to honor a foreigner, that I made many years later together with Walter Nernst, played out much more agreeably. This concerned the conferment of the golden Leibniz medal on Mr. Ernest Solvay in Brussels and at the same time on Mr. H. von Böttinger as recognition of their gains in the advancement of the sciences through donations of rich material means. Both gentlemen appeared in the Leibniz session on July 1, 1909 in order to accept this medal. In the evening, Nernst and I had the pleasure to organize a merry gathering in the rooms of the automobile club with many members of the Academy, with outstanding chemists and other scientific persons of Berlin to honor the two gentlemen. The task fell to me to greet the two gentlemen in an after-dinner speech. The celebration proceeded in harmonious and cheerful mood, and the speeches held there were collected in a small pamphlet: "Celebratory meal to honor the gentlemen E. Solvay and H. v. Böttinger on July 1, 1909 in the Imperial Automobile Club in Berlin". In appreciation of the honor Mr. von Böttinger made a donation of 30,000 marks, which he designated, on my advice, for the acquisition of mesothorium. I saw to the purchase of the preparation and held it until the Academy again moved into its new building *Unter den Linden* and thereby had the opportunity to store all material possessions at its own place. In the meantime, on my advice, the mesothorium was sold again by the Academy at a large profit, and a corresponding supply of the much more durable radium was procured.

Aside from the mesothorium, I also had to manage, for more than 20 years, the Academy's instrument collection, which was stored in my service apartment, at first in the Dorothy Street and later in the Hessian Street. Taking inventory of the individual objects was comparatively easy, but the traffic with the scholars to whom the instruments would be entrusted sometimes brought an uncomfortable exchange of letters. Also, I have in the meantime been rid of this duty, because the instrument collection likewise was transferred to the Academy's new building and is now managed by the body's archivist.

As the occupant of the academic-chemical laboratory and the associated service apartment in the Dorothy Street, one of the oldest scientific laboratories in Berlin, that was built by S. Marggraff and in which Achard for the first time produced sugar from sugar beets in larger quantities (1700 lb.), I stood in a special relationship with the Academy. This emerged particularly clearly, when the plan of a new building for the chemical institute was ready, and the transfer of the institute to a different location became necessary. For the approval of the construction money, the finance administration at that time set the condition that the Academy should give up ownership in the

Dorothy Street and would receive a share of the new building complex in the Hessian Street. It is understandable that giving up the old building, which belonged to the Academy for 150 years, and where not only the academic chemists but also, at times, the astronomer had found quarters, would be felt as painful by many a member of the body and that therefore the Academy's approval of the new plan would be made dependent on an exceptional negotiation. As the sentiment seemed to waver, two men staunchly for the plan stepped in, the medical doctor Rudolph Virchow and the theologian Adolf Harnack, with the remark that the Academy is duty-bound to come to the aid of a science like chemistry when it is in need. This generous opinion got through. The plan was unanimously approved, and I have always felt obliged to the Academy for this noble action. The liking that I had for it from the beginning was thereby greatly increased. Consequently, I have also accepted with particular pleasure an honor shown me by the Academy, when the Helmholtz medal was conferred on me in January 1909. This medal is not given according to statute, but usually every two years alternately to a physicist or a biologist. I received it according to the last characteristic, because my chemical work intersected biology many times. Through receipt of the medal, the current owner simultaneously receives the right to make suggestions for new conferrals, and I already made use of this to [nominate] van't Hoff to be so honored. He received the medal a few weeks before his death. It was the last honor conferred on him, and according to the testimony of the wife, it afforded him great joy.

Naturally, I always participated when a chemist was considered either for regular membership or as a corresponding member, and it is with great satisfaction that I can testify, that in every case involving chemists it was easy to reach an agreement. I want to make special mention of the selection of external members that took place at my suggestion: 1900 for M. Berthelot and 1905 for A. von Baeyer. Together with van't Hoff, I had the pleasure to announce this last selection at the celebration of Baeyer's 70th birthday in October 1905 in Munich.

Unfortunately, I could not take part at the celebration of the 200-year anniversary of the founding of the Academy, because I had to take an extended stay on the Riviera due to a tenacious weakness of the vocal cords. The celebration brought with it an increase in the membership, in particular, for the mathematical-natural science class, the division for technical sciences, in which as first members chosen were the gentlemen Müller-Breslau, professor at the technical college at Charlottenburg; Martens, director of the materials testing office at Lichterfelde; and von Hefner-Alteneck, electrotechnical inventor, previously a manager with the firm Siemens & Halske. The Academy thereby received the opportunity to honor outstanding external representatives of technology by naming [them] as corresponding members, and I have repeatedly allowed myself to suggest such choices for technical chemists, for example, for Ludwig Mond, Ernest Solvay, and Auer von Welsbach.

The importance of a scientific Academy is naturally determined first and foremost by the scientific reputation of the individual members. At the time of my entry, a comparatively large number of men of first rank belonged to the Berlin Academy. At the apex of the natural researchers stood Hermann von Helmholtz, a universal scientific genius; for he performed not only fundamental work in physics, physiology and mathematics, but also estimable accomplishments in epistemology. According to Alexander von Humboldt, he was the versatile natural researcher of the nineteenth century, not only in Germany, but probably in the world, and this versatility caused not the slightest break with the thoroughness of his research. In addition, he had, to a large degree, the gift of making scientific knowledge accessible to wider circles in easily understandable form and elegant language. I know few scientific works that in my younger years were so stimulating as the *Lectures on Various Branches of Physics, Physiology and Mathematics* published by Helmholtz. Perhaps Justus Liebig has had greater influence on the development of scientific and economic life in Germany through his popular writings on the importance of chemistry for agriculture, the arts, and commerce; but in subtlety of description and beauty of form they can't compare to the lectures of Helmholtz, to my taste. At the time when I got to know Helmholtz, he was already 72 years old and a sage personality in every aspect. For us younger [members] it was always a special pleasure when he rose in the faculty or the Academy to express his opinion in calm deliberate manner. I often had the opportunity to have private conversations with him in which he always expressed a kindly interest in chemistry. This personal and sincere contact between individual members of the Academy came about in informal manner at the post-sessions, which took place at coffee, first at the Rome Hotel and later at various other coffees *Unter den Linden* or in the Potsdam Street. These post-sessions were not infrequently more instructional and above all more entertaining than the official main session. When Helmholtz died in August 1894 it was a pleasant duty for me, as chair of the chemical society, of which he was an honorary member, to dedicate an obituary speech to him, which was printed in the report of the society, and in which I gave full expression to my admiration for the great man.

A second very interesting personality in the academic circle was Rudolph Virchow, pathological anatomist, hygienist, anthropologist, politician, and a man of unusual capacity for work. Despite the dissipation of his career, he treated everything he touched with great thoroughness and consideration. Even at the age of 80 years he typically dedicated not just the entire day, but also half the night to the work. He also had the talent to be able to use every spare minute, even in sessions or in company, for sleep. He was sharp in his judgment and could react most sharply to abuses of medicine and hygiene or against mistakes of the state administration. But I always had the impression that he was motivated only through objective grounds and high-minded political, social, or economic principles. To me personally he was always very friendly, and I have already [mentioned] his effective support for the chemical institute's new building. I remember a strong and successful protest he raised in the Academy against the acceptance in the session report of a treatise presented by S. Schwenderer of a certain Dr. Pinkus on the alleged connection between the growth of hair on the head and the fate of the wearer. His criticism was feared in biological circles but recognized by the majority as pertinent. He once took part at an evening meal I organized to honor William Ramsay. As we migrated from the warm dining room to the somewhat cooler living room, he jokingly said to me: "You seem, like the Shah of Persia, to have all climates at your disposal," whereupon I responded that in chemistry the period of extreme temperatures have begun.

Another outstanding medical doctor in the Academy was the physiologist E. Dubois-Reymond, [who] became particularly well known through the numerous academic speeches that he held as secretary of the body. He was doubtless also an erudite man and of unusual formal sophistication. From time to time even today I read individual speeches of his with pleasure, although it can't be denied that the grandstanding played a certain role in it, and the exaggerated mechanical tendency belongs to a different period of science.

Dubois-Reymond understood little of chemistry, but he had nevertheless been perceptive enough to devote to it a rather large division with independent leadership in his newly built physiological institute. When one remembers that E. Baumann and A. Kossel carried out a great part of their scientific work while chair of this division, one must concede to Dubois-Reymond that he had a lucky hand in his selection of co-workers.

The general lecture on the progress of the natural sciences, which [Dubois-Reymond] read in public every ten years was famous and gladly attended. Since his death, [the speech] has not been given, [and] it is doubtless a deficiency in the natural science instruction of the university. But at the moment, I know no one in our circle who would be [as] capable of this task as Dubois.

Among the younger natural researchers of the Academy, I met two old acquaintances, the botanist Engler, who had been *Privatdozent* in Munich at the same time as I, and the physicist August Kundt, my dear teacher from Strassburg, whom I have already mentioned earlier. In Berlin, he had launched himself with his own energy not only into the task of the teacher and researcher, but also into the vortex of society life. His health was not equal to it. A severe heart illness already at that time had become apparent, and in May 1894 he died of it. It was an agreeable duty for me to dedicate a brief obituary to him in the reports of the Chemical Society. Chemists will always thankfully recognize that, together with Warburg, he clearly demonstrated the monatomicity of mercury vapor by his general method of sound velocity and thereby made possible the same for the noble gases, discovered later. Kundt was an excellent teacher and dear fellow, who exercised a great force of attraction for younger natural researchers. His death ripped a large gap in our natural science circle. This was seen immediately in the appointment of his successor. At the suggestion of Helmholtz, the faculty presented only one candidate, Friedrich Kohlrausch, to the ministry. But before he accepted, Helmholtz died, and Kohlrausch became his successor at the physical-technical imperial institute in Charlottenburg.

The choice of the physicist at the university then, for the time being, had to be delayed, because no personality in Germany was available who was satisfactory to all involved. Over the course of the winter the thought emerged to win over J. H. van't Hoff for the post, after it became known that he would not be unhappy to leave Amsterdam. The instruction administration agreed quickly to this plan, and then Professor M. Planck was sent to Amsterdam to consider with van't Hoff the possibility of such an appointment. I have related the further course of the negotiation in a commemorative address for van't Hoff that I held soon after his death in the Leibniz session of the Academy. He declined the professorship at the university but was called to Berlin a year later and now received an associate professorship at the university with the honorary rank of professor, in order to be able to bring his salary to an appropriate level. Naturally I took part in van't Hoff's appointment and was thereby rewarded with a beautiful friendship that lasted until his death, and that brought me in close contact with this important natural researcher. Aside from a few small weaknesses he was a splendid man with ingenious mindset and scarcely less ingenious in his habits. He regularly held a public lecture at our institute on chemical theories while for his experimental studies he had, together with Meyerhofer, outfitted a modest private laboratory.

After van't Hoff had declined the professorship of physics at the university, Emil Warburg was appointed. Ten years later he went to the physical-technical imperial institute as successor to Kohlrausch, and Drude became professor of physics at the university but found a tragic end after a short time. Since then, Rubens, the discoverer of residual rays, has been the representative of experimental physics with us, while theoretical physics has been led and maintained by M. Planck in a most worthy manner during the entire time of my Berlin activity.

I had friendly relations with all of these physicists, who were, without exception, reasonable, benevolent, and calmly thinking individuals. This was particularly true of Friedrich Kohlrausch, with whom I was already friends in Würzburg. The casual, easy intercourse as Würzburg offered is scarcely possible in the big city, but nevertheless I came into very close personal contact with Kohlrausch through several experiments that we carried out together in the year 1898/99, with the goal of nothing less than a transmutation of the chemical elements. Long before such phenomena were observed with radioactive substances, I had imagined on the sun and other fixed stars, under the prevailing conditions of extreme temperatures and particularly pressure, that such transformations occur and that the true condition of the sun, where the chemical elements are concerned, is one of mass equilibrium. I have frequently expressed this view as hypothesis in my lectures, however never published, because I could not present any factual basis for it. I had then come to the thought, that perhaps elemental transformation could be achieved if one concentrate an enormous amount of energy upon a small quantity of material. I suggested to Kohlrausch to carry out such experiments with hydrogen in very rarified condition under extended treatment with cathode rays. The thought seemed not unreasonable, and we built ourselves a handsome apparatus which allowed us to concentrate the hydrogen treated with cathode rays at a pressure of 6–8 mm [of mercury] and then test spectrographically. In particular, we were hoping to transform the hydrogen entirely or partly into noble gas. The construction of the apparatus, the cathode rays, the vacuum pump, etc., fell naturally to the physicist. I assumed [responsibility] for the painstaking purification of the materials, the glass vessels, the mercury, and the hydrogen, which was always prepared finally from palladium hydride. We worked together many an afternoon until late into the evening, unfortunately without achieving a certain result. The cathode rays caused special difficulties, because with the purity of the cathode the resistance grew extraordinarily, and in the end such voltages had to be applied that the vessels could not withstand. We published nothing about this, and that is the reason I have spoken here about it in such detail. Perhaps 10 years later similar experiments were made public by English physicists, who were of the opinion that they had certainly observed the emergence of noble gas. However, they seem to have been mistaken; for later one heard nothing more of this discovery than disparaging criticism, and such an error can easily enter, as anyone who has tried similar experiments will agree.

Many years later I carried out another small project with Heinrich Rubens. This had to do with testing an assertion by the physiologist Rosenthal that alternating current can effect hydrolysis of starch and similar carbohydrates when the frequency reaches an appropriate but in every case different level. Our experiments [gave] entirely negative results but were not published because the matter no longer interested us as soon as Rosenthal's error was determined. Several years later a physical chemist also publicly refuted Rosenthal's assertion. The source of the error, however, remained hidden from him. I believe I found it through the observation that the starch paste used in fact can assume reducing characteristics when one subjects it to an extended treatment with alternating current in an open container; for then ozone forms in the surrounding air which quickly attacks the starch in the warmth and transforms it into reducing substances. One avoids this error completely as soon as one closes the starch paste containing vessel. The effect of ozone on starch paste, which, to my knowledge, is already well known in rough outline, deserves, by the way, a closer investigation in regard to the [reaction] products, and I hope to have the opportunity myself to acquire new experiences in it.

My special colleague both in the Academy and in the faculty was the chemist Hans Landolt, roughly 20 years older than I. As mentioned earlier, I could have been his successor at Aachen when he moved to the agricultural college in Berlin. From there he was appointed successor to Rammelsburg at the university and entrusted with the direction of the 2nd chemical institute of the university in the Bunsen Street. As an expert he was particularly strongly involved with my appointment at Berlin and also with entry into the Academy.

Landolt, the scion of an old Zürich family, did not disown his Swiss [heritage] despite his lengthy stay in Germany. He was a clever, critically inclined head and a man of sincere, pleasant character. He did not occupy an outstanding place as a researcher, but his books, particularly the large tables he published together with Börnstein, have enjoyed great esteem in the profession. We got along very well together and only extremely seldom represented different views in professional matters. He was not a leader, either in the science or in the Berlin body of scholars, but due to his prudent judgment and his general character he was popular in our circle. This was evident at the celebration of his 70th birthday, which he experienced shortly after a severe gall-stone operation. It was my pleasure to speak at this event during the celebratory meal and also to remember the fine collegial relation that continuously existed between us. I was also on friendly terms with his wife and the daughter, who was married to Professor O. Liebreich. Mrs. Landolt went above and beyond [in directing] the social traffic of chemical docents and assistants whom she frequently invited and thereby made possible for them the social connections which I unfortunately could not offer due to my withdrawn lifestyle.

At the age of 74 years Landolt gave up the professorship at the university and moved back to the physical-technical imperial institute at Charlottenburg, in order to conduct delicate experiments on the possible change of the total mass in chemical reactions, which he organized for more than 10 years at the 2nd chemical institute. He was particularly inclined for such experiments. It gave him great pleasure to make the weighings ever more precisely and to track down any possible source of error. After many iterations he came to the conclusion, as is well-known, that in chemical processes no change in weight detectable by current means [occurs.] This was nothing new; for this same law is a part of the fundamentals of chemistry, but nevertheless it is undoubtedly an achievement when even such fundamental laws from time to time are tested with advanced technical devices. The name Landolt will probably be tied more to this experiment in the history of the science than to his other scientific projects.

Walter Nernst, the outstanding physical chemist at Göttingen, became Landolt's successor at the university. At this time the 2nd chemical institute was rechristened "Physical Chemical Institute of the University" and the 1st chemical institute also lost its 1 and preserved the old simple name "Chemical Institute of the University."

Understandably, Nernst was also soon chosen for the Academy of Sciences, and chemistry was then represented by a group of 4 men, Landolt, van't Hoff, Nernst, and myself, who could assure suitable influence for chemistry in all relevant questions. Nernst, whom I already knew from Würzburg, because he was tested there by me in the doctoral exam, was later a dear and helpful colleague to me. In the faculty as well as in the Academy we almost always worked together as one. Only when we had to give advice to the culture minister on the occupation of the chemical chair at other Prussian universities did we occasionally differ. The sharpest [disagreement] came to light over the occupation of the professorship at Breslau after Buchner left. The faculty had proposed Abegg, and this candidate found enthusiastic support from Nernst. However, I was of the opinion that the professorship of experimental chemistry and the direction of the chemical institute should be assigned to an experimental chemist, while for Abegg a special professorship for physical chemistry should be created. After lengthy vacillation, the minister called Professor Biltz from Kiel to Breslau and thereby recognized the principle I stated as justified. Nernst seemed to have perceived this change of affairs as a setback not only for the person of Mr. Abegg, but also for physical chemistry, as became known to me indirectly only later. I want to acknowledge openly here that I never took steps to hinder the development of physical chemistry; [on the contrary,] I have always recognized it as a necessary part of instruction and research and also recommended [as much] to the authorities. I would even have arranged for a special division for physical chemistry in the new institute, had not the gentlemen Landolt and van't Hoff spoken against it on the grounds that they alone should be entrusted with the representation of physical chemistry. The justification of this wish was self-evident, and so I let the matter drop, although it would have been fairly easy at that time to win over Professor Bredig from Leipzig for our institute. My hope, that with the appointment of Nernst all chemistry students in Berlin would attend his lectures and exercises and be sufficiently instructed in physical chemistry, was, however, not fulfilled.

The big city brings with it [the expectation] that the student will limit himself as much as possible to the laboratory. With that also in the Berlin doctoral exam, chemical technology may be chosen as a minor, so that no room remains for physical chemistry. I hold, openly confessed, that this represents a defect in our instruction and I myself have always taken pains to make clear to students, during the administration of the doctoral exam, the importance of physical chemistry. But I can't hide the fact that it would be more effective to make it mandatory, and I hope this step will be taken after the war.

Nernst is not a narrowly focused expert; he takes interest also in industrial and economic matters, with which, through the invention of his electrical lamp, he found ample practical use. Further, he is an automobilist and farmer, hunter, and was even a soldier during the war. This is in accord with his lively intellect, his great energy, and his passionate opinion. Together with his charming wife and his children he has also established a hospitable house in Berlin, where the young chemists and physicists easily find inclusion. At the outset he was of the opinion that I might become envious of his social efforts and expressed himself frankly about it. My answer, that I placed not the least value in playing a social role and could only be thankful if he undertook this duty, was for him equally surprising and reassuring.

I will return later to our joint effort on the plan to build an imperial chemical institute and with the establishment of the Kaiser Wilhelm Institute for Chemistry. The same holds true for the gentlemen Haber, Beckmann, and Willstätter, who came to Berlin only after the establishment of the Kaiser Wilhelm Institute and then were accepted into the Academy.

I did not come into any close relationship with the minerologists and geologists. Nevertheless, I have made serious efforts to awaken interest in the young chemists for crystallography and minerology. I have also established a small number of minerals, for which I am indebted to the kindness of the gentleman colleague Klein, in an anteroom of the chemical institute. But I have not succeeded in achieving active participation of the chemists in crystallographic lectures or exercises. This has to do with the circumstances of the big city and the rules governing the conferring of doctoral degrees, which for chemists are not practical.

Among the biologists, I was mostly interested in the physiologists S. Schwendener, Haberlandt, and the medical doctor M. Rubner. All three were well-educated scholars and meritorious researchers. During the war I frequently met with the last two for advisory sessions on nutrition.

In the philosophical-historical class of the Academy at my entry, Theodor Mommsen was the most outstanding personality. In spite of the democratic constitution of the Academy, his influence in the philosophical-historical class was so great that he knew how to push through his opinion on nearly all important questions. For us natural researchers, his power became most apparent when he sought to block the selection of new members for his class. His powerful position, incidentally, was due to the unusual academic accomplishments and the extraordinary quick-witted criticism in which he did not disguise the polished and impassioned author of Roman history. Whoever has read this work cannot evade the impression of his genius and his enormous capacity for work. I myself had only short conversations with him here and there and provoked his astonishment with the information that as a *Gymnasiast* I had myself, at the encouragement of a senior teacher, prepared rubbings of Roman inscriptions from my homeland. For yet a second time I succeeded in moving him to a certain excited state when I led a difficult fight in the faculty against the doctoral candidacy of Russian chemists who lacked adequate preparation and thereby alluded to the earlier course of action by Mommsen against the certainly much worse mischief by several German universities of [allowing] doctoral candidacy *in absentia*. Consequently, he also moved to my side at the vote, and I thereby won a slim majority in the fight for the honor of my science.

At the Academy, I also got to know the historian Sybel, the author of the history of the French Revolution and the founding of the German empire, after I read his works in part. The last talk that he held for us was distinguished by elegance of form.

Treitschke came only later into the Academy, after Mommsen had died and other resistance was eliminated through the success of his German history. He seldom appeared in our circle, because he was already ill and died soon thereafter. However, I knew him in the faculty for a long time. At that time, he was already so hard of hearing that he could understand scarcely a word of the proceedings; nevertheless, his pride moved him to speak whenever possible, and for this purpose he had a neighbor write down key words that characterized the course of the debate. One evening I assumed this task myself, and I must confess, it was quite difficult to perform. When Treitschke correctly gathered the state of the proceeding from the key word, I was astonished to see with what speed he comprehended the entire complexity of the question and with his great eloquence expanded on it. It also happened, though, that he went entirely astray, due to [his] misunderstanding the key word, and then he talked about things that had no connection with the question at hand. The great passion with which he grasped and discussed, with poetic vivacity, everything interesting to him was characteristic. The strong influence that his lectures had with the academic youth was due in large part to this characteristic.

The Germanicist Erich Schmidt was of an entirely different nature. He was distinguished by physical beauty, resounding voice, and obliging manner. He was justifiably [respected] as an excellent speaker and preeminent authority of German literature. He was generously spoiled by Berlin society, and the nearly daily evening meals of which he partook year-round in the circle of high finance or upper-level officials contributed without doubt to his comparatively early death. One noticed with him not the slightest pedantry, found so frequently among philologists, even though he had a fundamental expertise and also remained active as a scholar. He showed great proficiency and assuredness in scholarly matters if they interested him. I have considered him the best deacon that the faculty has had during my 25-year membership. As his wife was born a Strecker and was half-sister to Mrs. Leube, I also came into social intercourse with him here and there, and I can only say, that like so many other people, I was taken in by his brilliant personality.

In a certain contrast to Erich Schmidt stood the classical philologist Diels, a splendid man, clever, benevolent, not narrowly focused in knowledge and outlook, and very useful as secretary of the Academy. Many speeches he held in this capacity were pearls marked by graceful form and thoughtful content. He had particular luck with his sons; for they all became acknowledged scholars, and one of them, Otto, was for years a dear pupil and coworker of mine.

The national economist C. Schmoller, at the same time well-known as a good historian, was also an interesting personality. When he addressed the Academy or the faculty on scholarly or business matters in his unassumed Swabian dialect, one could always be certain to hear clever and fully developed things. His special colleague Ad. Wagner, who, however, did not belong to the Academy, was much more vehement, and, despite being physically helpless at advanced age, always impressed us with his impassioned speech. Of course, I had to view his judgment in many things to be biased or eccentric.

Among the representatives of the humanities, I was drawn from the outset to one man in particular before I could know that I would later enter into a dedicated relationship with him in close joint work. That is the theologian and historian Adolf von Harnack. In the scholarly world he is rightly known as a polymath; for aside from his specialty, he has acquired excellent knowledge in philological, literary, and ethical areas, and in addition, he has succeeded in obtaining an overview of the natural sciences. As he often related to me, this was due, not so much to the marriage with a granddaughter of Liebig as much more to the nursery; for his three brothers studied natural science or medicine in Dorpat at nearly the same time as he. In this young company, who were physically very close to one another in the constrained apartment of the parents, it was easily understandable that a lively exchange of ideas would continuously take place. As a consequence of his outstanding memory, Adolf von Harnack was able to retain these impressions as an enduring possession. He first came closer to me personally on the death of my wife, whose family he knew well from a stay of several years in Erlangen, in particular, through his efforts to find a governess for my children. Later I was obliged to him for his support for the chemical institute's new building, which needed the approval of the Academy for this plan. He composed the history of the Academy for the 200-year anniversary celebration, which I rather diligently read, because this served as an excellent source for the history of the academic chemical laboratory. More than any other representative of the philosophical-historical class, Harnack always took pains also to promote the interests of the natural researchers, and he was therefore the man chosen to lead the Kaiser Wilhelm Society for the Advancement of Science when it was founded in the spring 1911. I will report on this later in detail. In the area of humanities, I hold Harnack at the moment to be the most outstanding German scholar, and as an organizer of scholarly work he surely earns the first position.

With the founding of the Kaiser Wilhelm Institute for Chemistry and Physical Chemistry, the circle of chemists naturally expanded, and among the [newcomers,] three men were chosen for the Academy, Beckmann, Haber, and Willstätter, about whom I will have more to say in connection with the Kaiser Wilhelm Society.

Chapter 10
The Chemical Society

At the time of my move to Berlin, scientific chemistry was represented mainly by the German Chemical Society, that Hofmann founded and led for 25 years in exemplary fashion and with great success. The celebration of its 25-year existence, already mentioned earlier, was the first scientific celebration that I experienced in my new position. Soon thereafter began the everyday work that I regularly took part in during the first 10 years, not only because I was repeatedly elected president and vice-president, but also because the Chemical Society at that time was housed as guest in our institute in the George Street. Their offices were namely the service apartment of the first assistant of the institute, the use of which earlier F. Tiemann and later S. Gabriel did without for the benefit of the Society. For the regular sessions, the auditorium of the institute served, and for all this the Society paid quite a small rent and a small recompense for lighting and heating. Also, most of the work for the business was honorarily performed by the president, the secretary, and the editor of the reports, so the expenditures of the Society, aside from the printing of the reports, were very small. It was thereby possible, through thrift over the course of perhaps 25 years, to amass a cash capital of 200,000 marks.

With the death of Hofmann, no disturbance in the business ensued, as F. Tiemann in usual fashion carried on in both offices as secretary and editor. At first, I held back and took part only in the sessions of the executive committee, in order to become familiar, as a neutral observer to the extent possible, with the ways of the business and the interested personnel. For the same reason, at the end of the year 1892, I also declined to be nominated as a candidate for president. Consequently, the leadership of the presidential business fell to Landolt. This was, of course, a mistake, as I later realized; for despite all of his good characteristics, Landolt was too careless. Also, due to his deafness, he was too clumsy in spirited debate to prevent a storm, which could have become too damaging to the Society. This was evoked through the opinion of several of the members of the executive committee who had been there since the founding of the Society, that Tiemann administered his offices too autocratically. The apparent ground was the proposal that M. Berthelot and C. Friedel in Paris be named honorary members. I had considered this together with Tiemann, and it was discussed

D. M. Behrman and E. J. Behrman, *Emil Fischer's "From My Life"*,
Springer Biographies, https://doi.org/10.1007/978-3-031-05156-2_10

in a session of the executive committee, at which, unfortunately, most members were absent, apparently because the naming of honorary members was not on the agenda. When it then went before the general assembly for a vote, the opposition, which had organized entirely in silence, fought the proposal, and the vote had to be postponed. This led to a tempestuous scene that the chairman Landolt could not control, and the session closed in quite unpleasant manner with a discord that seemed to signify the danger of a lasting split in the leadership. In the same session I was elected president and accepted the vote, but with a feeling of great unease and the resolve, under all circumstances to make similar scenes impossible.

I then had to spend several months doing rather difficult work behind the scenes to get the opposing sides, who were so strongly divided against each other in the leadership, to reconcile again and make it possible for Tiemann provisionally to remain as editor and secretary. For this it was necessary that I, as president, make myself entirely familiar with the substance of the administration, namely with the docket and the protocols of the scientific sessions and the sessions of the executive committee. This took place in the best accord with Tiemann, whose great capacity for work and solicitude for the business I have always acknowledged. A personal friendly relationship between us also developed, and when he suddenly died in the autumn 1899, I felt obliged to hold a memorial speech,[1] which was later supplemented on several points in a detailed necrologue by Witt. It is therefore not necessary that I dedicate still more words of friendly remembrance to him here.

One of our first joint tasks was the publicity for the election of Berthelot and Friedel, which had been postponed to the next regular general session. We felt obliged the more so, as both gentlemen had already been sent inquiries as to whether they would accept such an election. In order to eliminate the opposition of the small circle of Berlin colleagues, we turned to the large number of out-of-town members, who, without exception, agreed with the proposal. At the end of 1894, both French scholars together with Mendeleev and Beilstein were elected by an unusually large majority.

Soon thereafter the Society was set an entirely new and far-reaching task with the proposal of Mr. F. Beilstein to transfer to [the Society] all his rights to the *Handbook of Organic Chemistry*, which had recently appeared in its 3rd edition, if [the Society] would agree to allow supplemental volumes to appear and later to organize a new edition. At that time Beilstein already felt he was too old to accomplish this large task by himself. He had first turned to Mr. Paul Jacobson with the inquiry as to whether he wanted to assume the editing of the book. Although Jacobson was particularly well prepared, through the publishing of the well-known outstanding textbook of organic chemistry by Victor Meyer and [himself,] it seemed to him that the new task was too difficult for a single person, and he made the suggestion that the Society should assume the [responsibility] and expand [the book] further through the establishment of a central office for the delivery of chemical reports. At the same time, he declared himself ready, under suitable conditions, to assume [leadership] of such an office.

With that, a thought took practical form that I had long had in mind, and which I had repeatedly discussed with Tiemann and other colleagues in small circles: the

---

[1] Reports of the German Chemical Society, vol. 32, p. 3239.

centralization of the report delivery for scientific chemistry, which until then had been strongly scattered, and thereby making possible a reduction in cost and a wider circulation for the original publications. Until then there was the long-standing *Jahresbericht fur Chemie*, founded by Liebig and Wöhler, which had regularly appeared since 1848 and was tied directly to Berzelius's yearly report. Also in existence was the weekly *Zentralblatt*, founded by Arendt, wherein each individual publication handled things differently, resulting in the literature of chemists not being conveyed systematically, [though] in a short time. The reports of the Chemical Society were somewhat similar, though in much more fragmentary form, in the papers section. Finally, there were still other papers in the magazine of the Association for the Protection of Interests of the Chemical Industry, the *Chemist* newspaper, the *Apothecary* newspaper, and similar organs. It was obvious that this fragmentation of report delivery was inexpedient and that with the same resources, much more completeness could be accomplished.

Tiemann was of the same view, and so in 1895 we both made the decision to implement Jacobson's suggestion and procure the assent of the executive committee and later of the general assembly. This succeeded, not without considerable trouble and worry. Already in the executive committee, several careful members, for example, the otherwise so benevolent C. Liebermann, announced against the plan, because they foresaw in the new book business a danger for the prosperity and particularly for the finances of the Society. On the other hand, we were able to determine, by calculation, that the Chemical Society would be in the position to win over a much larger number of subscribers for the chemical *Zentralblatt* and for the Beilstein handbook and thereby make possible a considerable reduction in price. No one could deny that this would be a victory for our science and also must enhance the reputation of the Society. Thereupon the number in favor quickly grew, and in the general assembly on December 13, 1895, I was able, as chairman, to make the first public announcement about it. The decision came in an extraordinary session of the general assembly on June 19, 1896. Under the leadership of Dr. Jaffe, the opposition again spoke sharply at the decision. But as chairman, I was able to refute all complaints and fears, and in the end, the change in statute necessary for the plan was approved by a large majority.

First of all, it had become necessary to purchase the publishing rights to both works from the book publisher, the firm Leopold Voss in Hamburg. I had already taken the decisive step, together with the treasurer Dr. Holtz, who from the outset was in favor of the plan and energetically promoted it on every occasion, in the autumn vacation 1895. In conjunction with a meeting of the Association for the Protection of Interests of the Chemical Industry in Kiel, in which I took part, I drove with Dr. Holtz to Hamburg and agreed with the leader of the firm L. Voss a fixed price of 15,000 marks for the *Zentralblatt* and for the supplemental volumes of the Beilstein a quite favorable share of profits for the firm. When [the firm] later hesitated to recognize the verbal agreement in writing, and we were concerned that the [signed] contract could not be presented in time to the general assembly in June 1896, I advised Tiemann to bring heavier guns to bear and the matter to a close. He succeeded in bringing the reluctant firm to reason through highest-energy telegrams.

Meanwhile, negotiations had been maintained between Paul Jacobson and the previous editor of the chemical *Zentralblatt*, Professor R. Arendt in Leipzig. Both entered the service of the Chemical Society at the start of 1897. Arendt remained in Leipzig and saw to the editing of the chemical *Zentralblatt* there until his death. Jacobson moved to Berlin and assumed not only the editing of the Beilstein handbook, but also relieved Tiemann as general secretary of the Society in the editing of the reports and administration of the business of the secretary. Tiemann thus left both offices that he had managed unpaid for nearly 20 years, and the executive committee felt obliged to express their warmest thanks by presenting him with a laudatory address and by organizing a celebratory meal.

This new order of the business leadership with the expanded literary tasks signaled the start of a new period in the development of the Chemical Society. That the change was correct is indicated by success. To the four volumes of the third edition of the Beilstein handbook the Society has published an equal number of supplemental volumes and prepared a new edition, which has a value of not less than 1.2 million [marks] and a compass that will attain encyclopedic [proportions.]

Also, the chemical *Zentralblatt* has developed in gratifying manner, first under the leadership of Arendt and then, after [Arendt's] death, under the leadership of Professor A. Hesse. The number of subscribers has more than tripled, the compass has continuously increased and also the quality of the individual papers has improved. Only through the war has a decline in all of these points occurred. In addition, the Chemical Society has taken over two new works: the *Literature Registry of Organic Chemistry*, founded as a lexicon for organic compounds by M. M. Richter, and now carried on by Professor Stelzner; then a similar literature registry of inorganic compounds founded and led by Dr. M. K. Hoffmann.

The importance and usefulness of these different enterprises will not be denied by any German chemist, and the chemical industry later expressed its agreement through rich material support. They began in 1909 with a donation of 60,000 marks from the firm Cassella & Co. in Frankfurt-on-the-Main. There followed soon after a collection of 200,000 marks as Beilstein foundation, started by the new treasurer Dr. F. Oppenheim. Then came 2 collections for the lexicon of inorganic compounds and for the literature registry for organic compounds in the amounts of 75,000 marks and 120,000 marks. Finally, during the war, the Chemical Society, at its 50-year jubilee, was given the gigantic sum of 2½ million marks by the industry and several private [individuals] as a guarantee for its literary efforts.

At the same time, an entire staff of scientific civil servants was formed, primarily for these literary matters. The Society's house, named after the founder Hofmann, also assumed thereby a worthy and serious purpose. The plan for its foundation was conceived immediately after Hofmann's death and also found manifold support in the circle of the out-of-town members. I myself was taken with this plan from the beginning and also demonstrated [my support] with a contribution (2000 marks) that was rather high for my circumstances at that time. But when I came to Berlin and was elected to the commission for the establishment of the house, I was shocked with the irresponsible manner, as I saw it, in which the project was to be financed. The collection had amounted to only about 200,000 marks, and now a house was to be erected estimated to cost, including the building site, 800,000 to 1 million marks. This would use not only the entire savings of the Society but would also [result] in a considerable mortgage. I felt obligated to raise energetic objection against this, and I also saw to it that the majority of members of the executive committee came to the same view, that the Hofmann House must be built by other means and passed on to the Society free of charge. The building was thereby surely delayed, but there was no harm it that, since the Society had in our institute a modest, to be sure, but sufficient home. At the same time, the adherents of the luxurious building were compelled to find a different form of finance. This succeeded, due to the efforts of the gentlemen Holtz, Krämer, Martius, and Tiemann, when they formed a limited partnership for the building of the house, whose members' shares were [each] at least 5000 marks. Originally an interest payment for [the shares] was in prospect, but later the gentlemen managed to arrange for most of the owners of the shares to renounce any indemnity or repayment of the sum. I myself took no part in this business. Also, the building of the Hofmann House occurred without my special collaboration, only I was allowed, in keeping with a wish of Mr. Jacobson and the other scientific civil servants, to outfit a small laboratory whose equipment could also be used for experimental [demonstrations] in the large session room. The house was opened 1900 with a celebratory session under the leadership of the president J. Volhard, and A. v. Baeyer gave the main talk on the history of indigo.

The creation of the Hofmann House is primarily to the credit of J. F. Holtz, who managed the collection of the money with vigor and patience. However, he also took the blame that the construction turned out quite costly and that the original building costs were overrun by about 80,000 marks. The interest for a mortgage loan of 100,000 marks and the not insignificant maintenance costs naturally fell to the Society as a burden, so that the house cost [the Society] yearly 12,000–15,000 marks. I did not hesitate to raise an objection to the sharp increase in expenses of the Chemical Society due essentially to the Hofmann House, and thereby it came occasionally to ill feeling between Holtz and myself. However, since the finances of the Society were shaping up favorably in the new century despite the expanded literary enterprises, and with the growing number of civil servants the building was always being better utilized, I was content and am now reassured about the financial fate of the Society. For reasons of thrift, the upper part of the building was rented in the first years to the Professional Association of Chemists. Since then, though, they have occupied their own home next to the Hofmann House, [and] the upper rooms have also been used only by the Chemical Society. Of course, the service apartment of Professor Jacobson occupies a part of [this space,] which was also allowed to him, considering his services to the Chemical Society, after he gave up the greater part of his duties several months ago to [focus] on his textbook. This service apartment represents a reserve of space, which will surely be enlisted later for business tasks of the Society.

The auditorium of the Hofmann House is decorated with numerous pictures of deceased chemists. I donated the [picture] of Georg Ernst Stahl, the originator of the phlogiston theory. It is a copy by Walter Miehe of the original in possession of the Kaiser Wilhelm Academy for Military Medicinal Science. The picture shows the outstanding medical doctor and chemist, who played a large role in Berlin as personal physician to King Frederick William I, in the traditional costume of the eighteenth century with velvet coat, lace collar, and a large full-bottomed wig. A second copy of this picture that I [commissioned] two years ago by Miss Chales de Beaulieu and was designated for the German Museum of Munich had a less favorable outcome, because it was declared not suitable for the museum by Mr. Oscar von Miller, and [it] is still in my possession.

During the first 8 years of my stay in Berlin, I took part almost without exception in the scientific sessions of the Chemical Society, for we were under the same roof, and I also presented a large part of my scientific experiments there. The first report in January 1893 concerned the discovery of aminoacetaldehyde. My suggestion, to test the physiological effects of the chloride salt, caused a small debate in the session. However, it was easy for me to lead the opposition to the absurd with the remark that plants, to my knowledge, are living things with no nerves, and one must deal with all physiological questions without preconceived opinions. From then on, the younger chemists in Berlin became somewhat more careful when they criticized my reports in the Society, although I was always ready to recognize every sensible and properly presented opposition and deal [with it] objectively. The sessions at that time were occupied almost exclusively with original talks, which in many cases were elaborated in exaggerated form and became boring, while the numerous communications sent in from out-of-town, whose content, as a rule, was far more interesting, were dispatched by the secretary in a few words. At my suggestion, the change was made in the year 1896, so that the out-of-town communications would be presented by a larger number of young chemists in greatly detailed manner, whereby the sessions doubtless gained in content and interest.

On the other hand, the executive committee took pains, with full justification, to hold summary lectures on broader fields of work by the best experts, which had already been established several years before Hofmann's death. I participated fairly regularly in selecting the speakers the first 26 years. Particularly memorable to me were the lectures of van't Hoff on the new theory of solutions and of Ramsay on the noble gases. Van't Hoff came to Berlin for the first time for the occasion. After the session he was honored at a small festive meal, and I celebrated him as a king of the science in connection with his name "von Hofe." The impression that he left with us at the time was not entirely without influence on his later appointment at Berlin.

Ramsay was celebrated even more; for the public at large took part in his discovery, and he was able to repeat his lecture in a modified popular form on the one hand before the royal couple in the auditorium of the chemical institute and on the other at Urania. At the lecture for the Kaiser, only the presidency of the Chemical Society took part, and it was the first time that I came in contact with Wilhelm II and the Empress, to be sure, only briefly. Professor Slaby communicated to us the Kaiser's wish to hear a lecture by Ramsay, and I undertook to ask Ramsay if he would be willing. The laconic answer "Yes" was characteristic of Englishmen. Ramsay was my guest during his stay in Berlin, and I became closer to him personally during these days, where we were able to chat for hours.

He was not only an excellent natural researcher, but also a generally refined, very eloquent, clever, and deliberate man. He [visited] me a second time, at the new service apartment Hessen Street 2, during the International Chemists Congress 1903, and the favorable impression from earlier was thereby only strengthened. I later visited him again in London and have remained in fairly regular correspondence with him. In particular, he reported to me several times on his radioactive research. The last letter, that I received 6 weeks before the outbreak of the world war, was delivered in an unusual tone and gave tidings of the vehemence with which he apprehended political questions. At that time, he was most agitated by the politics of home rule of the English government in Ireland and stood on the side of the Ulster Party, which in his opinion must resist home rule with armed force, and for which he had decided personally to provide assistance. This was the forerunner of the extreme attack, caused by the war, that he launched soon thereafter against Germany, and which I will not go into for the time being.

On his first visit in December 1908, I organized for him an evening company for gentlemen, at which several Berlin chemists and a larger number of academics were invited. Ramsay had forgotten the time and came home therefore only after the entire company had long gathered. I then could only admire the speed with which he changed from the street suit to evening dress. I believe the entire transformation took less than 3 min.

A larger celebratory meal had been given for him days before by the Chemical Society, and I had to make the after-dinner speech to the guest, which I was able to clothe in humorous form. It was printed, and I had several copies sent to outstanding out-of-town colleagues. The most remarkable effect was in Munich; for Pettenkofer, whom I had sent the sheet to, sent it on to W. von Miller, who at that time was already seriously ill and whom he wanted to cheer up a little. Miller, who suffered from colon cancer, had the personal pride to lecture on chemistry at the technical college with the reserves of energy that his depleted condition would allow. For this purpose, he had himself carried into the lecture hall on a stretcher. He completed one of his last lectures mainly by reading my after-dinner speech, while tying in, of course, instructive remarks about the noble gases and their discovery. That brought in various letters to me from his audience who were very amused by it.

Other noteworthy talks were held by Nernst, Wallach, Buchner, Willstätter, Richards, Sabatier, Brunck, Knietsch, etc. Everyone listened with lively interest and I, acting as chairman from time to time, extended to the speaker the thanks of the Society. R. Fittig held a talk in a peculiar style on unsaturated [fatty] acids, lactones, etc., very rich in content, but so peculiar in form that the youth were very amused and we older [men] had to take pains to remain serious.

My first talk on carbohydrates, which followed those by Baeyer and Victor Meyer, was already mentioned previously. I gave a second talk in January 1906 on amino acids, polypeptides, and proteins. It created somewhat of a stir among the colleagues. Perhaps 8 days after the talk the attention of the press and public at large was aroused through an item from Vienna, and I experienced something very curious. My carefully [prepared] talk was embellished by the press in the most fantastic manner: the synthesis of egg white represented as a finished work and proclaimed as a solution to the age-old question of nutrition. I tried to raise objection against this but without any success. Like an ocean wave, the flood of newspaper articles washed over me and all reason. In the end I received a newspaper article from the United States of America, covering the entire page and illustrated with pictures, in which one could see the transformation of coal into the most beautiful products of an elegant restaurant. Naturally all of these pictures of myself, my laboratory, the experimental animals, etc., had been freely invented. At that time, I got a fright from the terrible power and lack of restraint of the press.

I held the third summary lecture in September 1913 in a session organized by the Chemical Society for the meeting of natural researchers in Vienna. The theme was the syntheses of depsides, lichen materials, and tanning materials. I was assisted in the talk by my son Hermann O. L. Fischer, who had taken part experimentally in the syntheses of the lichen materials. The session took place in the auditorium of the chemical institute of the University of Vienna, which at that time was led by Guido Goldschmidt. The room was so overfilled that no space remained for me behind the experiment table to elucidate the numerous charts and to demonstrate the preparations. Also, it was so hot that after the 1½ h talk I was as wet as if I had swum across the Danube. It was the first special session which the Chemical Society held at a meeting of the German natural researchers and until now the only one, because due to the war, no meeting of the natural researchers has come about.

When the chemical institute moved into the new building in the Hessen Street in the year 1900, the nearly daily contact between the Chemical Society and myself was interrupted. At my request, the Society received permission from the culture ministry to remain in their old space for another year, until the Hofmann House was entirely finished. But with the physical separation I had the feeling that I would no longer be responsible for the fate of the Society to the same degree as before; rather, that the concern would be more equally divided among all members of the executive committee. In keeping with this, I only seldom again accepted the vote of the executive committee [to make me chair] and also took part in the Society's sessions in freer form than had been the case earlier. Only when important matters were in play did I not shy away from exerting my entire influence to push through that which I held to be right.

I could not attend the 50-year anniversary of the Society that was celebrated in May of the same year,[2] because I was in southern Switzerland at the time, recovering from a lung inflammation.

I had placed a request beforehand with the ministry of culture, however, to give public recognition to the scientific civil servants of the Society, with Mr. P. Jacobson at the head, by conferring titles. This in fact occurred, and to my great surprise I was also granted a high Prussian honor at the occasion. To be honest I would much rather that this had not occurred.

---

[2] Translator's note. The Society was founded in 1867, so Fischer seems to refer here to the year 1917.

Unfortunately, I had to decline to give the memorial speech for A. von Baeyer that I [had] proposed and originally assumed, due to illness. It was given by R. Willstätter in very worthy manner.

At the 40-year anniversary of the Society a small celebration was also organized, and I had the pleasure on this occasion to have Mr. Professor Henry Armstrong, sent from the Chemical Society of London, as my guest. I have long had friendly relations with him, partly through his son, who was my co-worker for several years, and also through my own son Hermann, to whom he and his family extended friendly [hospitality] during a half-year stay in England.

Chapter 11
The Chemical Institute in the George Street

My move into the institute, the small structural changes suggested by me, the increase in the number of assistants, and the division of the work among the available personnel have already been portrayed earlier. All of this was thought to be provisional, because I was certainly counting on [the hope] that a new building would arise in a few years; for this [new building] was the first and most important stipulation I set at my appointment, and this was agreed to not only by the councilors at the culture ministry but also by the minister Dr. Bosse himself. The provisional [period,] however, lasted 7½ years and so formed a rather long portion of my activity in Berlin.

With the exception of the two lecture halls and both rooms used by Tiemann as a private laboratory, all the usable rooms were on the first floor.[1] Two well-lit classrooms, separated by a smaller room, faced the George Street. The remaining classrooms with anterooms lay in the wings, which resembled corridors, and served also as passageways.

The private laboratory consisting of two rooms was quite handsome and also rather suitably furnished. It stood through the corridor and two library rooms, directly connected to the dwelling house in the Dorothy Street. Supported by the private assistant Dr. Lorenz Ach, who came along from Würzburg, I quickly arranged myself. Three young chemists who wanted to carry out their doctoral work—it was the gentlemen Haenisch, Kopisch, and the Englishman Crossley, who had come along from Würzburg—were accommodated as guests. Already in the first year I had the good fortune to find two pretty reactions: the preparation of amino-acetaldehyde and similar materials, further, the synthesis of the alcohol glucoside from sugar and alcohol in the presence of hydrochloric acid. At the same time, I was leading several doctoral projects in the second organic classroom and, together with the assistants occupied here, overseeing the preparative exercises of the organic beginners following the book I wrote: "Introduction to the Synthesis of Organic Preparations," which had already proved itself in Erlangen and Würzburg. I was also involved with

---

[1] Translator's note: in German parlance, the "first floor" of a building is the first floor *above* the ground floor.

© The Author(s), under exclusive license to Springer Nature Switzerland AG 2022
D. M. Behrman and E. J. Behrman, *Emil Fischer's "From My Life"*,
Springer Biographies, https://doi.org/10.1007/978-3-031-05156-2_11

the analytic division, in which Dr. Piloty and Dr. Fogh were occupied, to the extent that my time permitted.

Like Hofmann, I held the large experimental lectures in the winter on inorganic and in the summer on organic chemistry. However, instead of two-hour lectures three days per week, I preferred to speak 5 times per week for one hour, because the preparations of the experiments become much more careful thereby and also quite a bit of time can be saved. In fact, it seemed that I could also cover a larger syllabus with a greater number of experiments in the five weekly hours than previously was accomplished in the three double-hours.

The number of listeners grew considerably already in the second and third years, and then the new registrations became so large that all available space was occupied by standing persons, and most of the entry doors could no longer be closed. The cloak room was entirely insufficient. Frequently at the end of the lecture there was great disorder here, and unfortunately also some theft occurred.

Among the listeners there were in the last years not only chemists, medical students, apothecaries and teacher candidates, but also members of the law and even the theological faculty. This may well have to do with the fact that I tried occasionally to explain the importance of chemistry for the entire human culture in the philosophical and economic connections.

Also, the laboratory was soon overflowing, which could not have been surprising, given the limited number of available seats (about 80.) This rush was the best incentive for the new building for the institute, and when I pressed for it, I received the directive, already in the spring 1893, to elaborate a program for it. I immediately undertook this task in a joyful mood. However, to my disappointment, I soon had to recognize that the directive was given as a mere pretense, and that a much heavier gun would be needed to bring the culture ministry to consider this rather difficult question energetically. The finance minister has the final decision for all such requests of science in Prussia that cost money, and I will later describe how difficult it was to obtain his agreement.

Prior to that, the culture ministry foisted on me a rather uncomfortable business that was connected to the World's Fair of 1893 in Chicago. The ministry had undertaken to organize an exhibit of the manner of Prussian teaching and scientific research in Germany. The Chemical Society was to participate, and at the request of the ministry I made efforts to obtain the approval of the executive committee. Above all, [the ministry] wanted an exhibit of interesting preparations discovered in Germany. At the outset, the executive committee showed no inclination to agree to this. I [then] made clear to the gentlemen that use of space in the institute for purposes of the Society, which had gone on for years, was possible only with the assent of the minister and that future use thereof would also be dependent on this assent. Only then was it decided to cooperate for Chicago through organization of a collection of interesting preparations and submission of an album with pictures of German chemists. I was obliged, however, to design and send off the circulars to the members of the Society with the request for participation as well as to take receipt of the preparations. This work was done in the rooms of the institute with the help of the assistant Dr. Richter. It took weeks of aggravating effort and much worry. A handsome display case was obtained gratis from the publicly traded firm

Chemical Factory, previously E. Schering, in Berlin. The entire [exhibit] later in Chicago apparently was pleasing; for the participating laboratories were rewarded with certificates of honor; also, we were met with a wish of the Americans to be allowed to retain the collection. The Society could not accede, however, since the preparations had remained private property of the individual members, and nearly all of them had expressed the wish to get them back. After the close of the Chicago fair, they were then returned undamaged to Berlin and from here distributed to the individual owners. I had so much inconvenience from the entire business, however, that on later impositions of a similar nature on the part of the culture ministry, I energetically declined. Consequently, the organization and participation of the scientific chemistry for the [World's] Fair in Philadelphia was assigned by the culture ministry to Mr. C. Harries.

The first winter that I spent in Berlin was unusually severe. At that time, I did not yet have the service apartment, but rather lived in the Queen Augusta Street, so I had to make the journey 4–6 times per day, naturally with a wagon, which, however, due to the high snow, took nearly ½ hour. At the end of the journey in the severe cold I was entirely frozen to death, until my wife came to the fortunate thought to buy me a large fur muff into which I slipped my arms during the wagon drive.

The frost also had very unpleasant consequences in the laboratory, for the water pipes burst in various places, and there ensued a great deluge. Unfortunately, the pipes were all laid inside the walls, and no plan existed anymore for the layout. Consequently, the construction crew often had to search all day to find the damaged place. My earlier resolve, to leave the pipes free in the new building to make them easily accessible, was thereby strengthened. By the way, despite these small and unpleasant surprises, the instruction and scientific work in the institute went forward undisturbed.

After Mrs. von Hofmann, in May 1893, left the service apartment in the Dorothy Street and a very necessary repair job on it was undertaken, I was able in September of the same year to take possession and was, from then on, also in a position, not only to engage more intensively in experimental work, but also to manage oversight of the entire institute more carefully. This soon led to the discovery of theft committed or attempted by the depraved custodian of the institute. He had taken and sold a set of copper plate from the instrument collection of the Academy, which was located in an attic room of the living quarters. I became aware of this when I assumed supervision of the instrument collection on moving into the service apartment. Also, during the time [the apartment was vacant,] he had removed the copper kettle embedded in the wall of the washroom, in order to sell it, but was surprised by the construction supervisor. Consequently, I gave him notice, and then the fellow had the effrontery to complain about [ill] treatment allotted him by the new director of the institute in a petition to the culture ministry. This was conveyed to me in a report from the minister, and when I replied that the reason for firing [him] was compelling suspicion of theft, there came the prompt inquiry, why no [criminal] charge should follow. I had foregone this measure because I felt sorry for the culprit; for he had come to the disordered lifestyle and moral turpitude through lack of supervision and also through the bad, damp, and dark cellar apartment. His successor, a coppersmith [named] Prisemuth, who through the wearing of eyeglasses had an entirely scholarly appearance, was not much better. First, I had an uncomfortable scene with him because of his wife, with whom he frequently quarreled. One Saturday afternoon I was called home from a medical student's qualifying exam, which took place in the physiological institute of the university, because the custodian's wife had supposedly had an accident. When I appeared at the institute in the George Street, the custodian was calmly engaged in cleaning the corridors. I excitedly inquired how things stood with his wife and received the response: "Stupid jealousy, Herr Professor, she has already made me the fool several times." In reality, the wife had attempted suicide by drinking strong hydrochloric acid and thereby suffered severe burns of the mouth, voice box, and throat. The Mr. Husband showed not the slightest sympathy for [her,] and when the wife returned to him from the hospital several months later, he was apparently entirely pleased that she had lost her voice and could only whisper. That was an internal family matter to which I was not entitled to apply any measure. But 1½ years later I learned that the custodian, despite strict prohibition, was renting out master keys to students, so there was nothing else to do but toss him to the wind, even though he was quite a skillful worker. He was still retained and valued as a servant by Tiemann for his private laboratory until various misdeeds made his dismissal necessary here also.

Another servant, the oldest in the building, proved to be just as unreliable. He lied in the most unashamed manner, even when one could prove his guilt. He also had the reputation of being unreliable as the caretaker of the glass apparatus and chemicals. When he was 65 years old, therefore, I requested his retirement, but it was quite difficult to achieve this, even though legal grounds [for retirement] for age were in place. In the end he was mollified by the conferment of a general certificate of honor, and I saw by this example what great power the government authorities have, even with the lower officials, in the conferment of certificates of honor and orders [of merit.] The inferior quality of the old servant I believed at least in part due to the poor condition of the service apartment. For the new building I therefore took care that the apartments of all lower-ranking civil servants be equipped hygienically entirely flawlessly and also comfortably furnished with lighting, heating and the like. Since then, I have not experienced such sad things as previously described, and only during the war, with its demoralizing influence, have relations with servants again taken several uncomfortable steps. I therefore gladly agree with the judgment of discerning political economists, that a good apartment improves people and a bad one oppresses them ethically.

Almost without exception I got along well with the assistants of the institute, whose number grew gradually with helpers and private assistants. They were always appointed by the minister for only two years, but this appointment was renewed as often as needed at my request for those gentlemen who wanted to dedicate themselves to a scientific career. The wish to renew an appointment was rejected, to my knowledge, only a single time, because the individual in question had behaved improperly toward the minister himself.

From the outset S. Gabriel stood at the summit of the assistants. He had already been with the institute under Hofmann for 20 years, first as student and then as assistant. He has always been a dear colleague and friend. I therefore gladly take this opportunity to express my heartfelt thanks for the steady, friendly manner of help that he provided so often when [I] was sick or indisposed as well as in the lectures and in the administration of the institute. As I was also convinced of his scientific excellence, I would have gladly helped him to obtain an independent livelihood, but despite all recommendations, [we] did not succeed in getting him an appointment at another college. Only with the new institute was it possible to give him a permanent appointment as head of a division, and I later also arranged that his service around the institute and in instruction be recognized through conferment of orders, the title of *Geheimer Regierungsrat* and naming as honorary professor in the faculty. He is a year older [than I] but considering his good health I have every reason to hope that he remain active at the institute at least as long as I.

The two other assistants that I inherited from the Hofmann time, Dr. Richter and Dr. Pulvermacher, left this position voluntarily after the elapse of 1–3 years. I have never heard anything more from the first. The second was secretary general of the International Chemical Congress in Berlin perhaps 10 years later. Dr. C. Harries remained much longer. Already with Hofmann he was a provisional lecture assistant. He received a regular assistantship in the same capacity on my entrance, later became instructional assistant, and in the year 1900, at the opening of the new institute, division head for organic chemistry. After the retirement of Claisen he was appointed [full] professor and director of the institute at Kiel. He gave up this position 1½ years ago and moved back to Berlin-Grunewald as a private scholar. Harries performed his doctoral work under Tiemann, but he began his independent experiments in my time, and with the ongoing personal and scientific discourse between us for years, he learned so much from me that he might also well be counted among my pupils. In fact, one of his first projects concerned also the so-called phenyl-sulfocarbazin, discovered by Besthorn and myself, from which he established that it did not have the structure supposed by us, but rather that the ring contains the sulfur. However, he also did not succeed in the final identification of his structure.[2]

Harries began his experiments on the oxidation of unsaturated bodies with ozone at the Berlin institute, which he later extended to india rubber and thereby achieved his great scientific success.

The two instructional assistants brought with me from Würzburg, Dr. Piloty and Dr. Fogh, experienced different fates. The first I have already discussed in detail. He carried out his beautiful experiments on dihydroxyacetone, on a new synthesis of glycerine and a special class of nitroso- compounds. His wife, Eugenie, daughter of Baeyer, had already befriended my wife in Würzburg. Her first child, a son, was born in Berlin Christmas 1892, and I had the honor to be chosen as godfather. Like the father, he unfortunately fell in the unholy war in the west.

Dr. Fogh, a Dane, had little success in Berlin. He entirely laid aside the thermochemical experiments begun in Paris with Berthelot and continued in Würzburg. After several semesters he also became ill, first took leave, and, since [his condition]

---

[2] Translator's note: This compound is currently known as 1-thiocarbonyl-2-phenylhydrazine, which does *not* have sulfur in the benzene ring.

did not improve, voluntarily gave up his position. He moved back to Copenhagen, and aside from an engagement notice, I have received no more news from him.

The private assistant Dr. Lorenz Ach, who came with me from Würzburg, had much better success. He provided valuable help in my investigation of aminoacetaldehyde and the alcohol glucoside. After this activity he became instructional assistant in the organic division, and I suggested to him a joint project on the transformation of 1,3-dimethylpseudo uric acid, which shortly before I had had a student prepare, into the corresponding uric acid. In fact, the reaction succeeded first through fusion with oxalic acid, which also proved usable for the analogous synthesis of uric acid. Following this experiment came the first synthesis of caffeine. Before this, however, Dr. Ach moved into engineering science, indeed, to the scientific laboratory of the firm C. F. Böhringer and Sons in Mannheim-Waldhof, whose leadership was entrusted to his brother, the previously-mentioned Dr. Fritz Ach. After the untimely death of the brother, Lorenz became the head of the laboratory and achieved handsome success here.

Through these personal connections I was also prevailed upon to entrust this firm with the technical preparation for the synthesis of caffeine. I will later discuss the project that developed from that in connection with the general investigations on purine.

Dr. Rehländer, an equally splendid, thorough, and diligent chemist, became Dr. Ach's successor in the private laboratory. He took part, not only with the preparation of the glucoside synthesis, but also played an active role in the extensive experiments on enzymes, which began with the characteristic behavior of the α- and β-glucosides toward emulsin. He later joined the von Schering factory. At the same time Dr. Beensch acted as private assistant and then went, like the brothers Ach, to the firm Böhringer & Sons. He was replaced in the private laboratory by P. Hunsalz, who remained with me longer than usual and participated to outstanding degree in the synthesis of purine.

[Hunsalz] was a native of Memel, stemmed from modest circumstances and was physically an unattractive man, but clever and extraordinarily diligent. I had already struggled energetically, with success, against his inclination to hurry and to rush into the experiments during his doctoral work on the hydrazinoaldehyde. He was later quite a skilled experimentalist who was particularly good at very quick work. He was the first private assistant whom I granted, considering his accomplishments, quite a considerable subsidy to the state salary. In the end he also came to the laboratory of Böhringer & Sons, but did not feel at ease here, apparently due to the predominance of the brothers Ach, and therefore went, after several years, to the firm Schering in Berlin. In hindsight this decision seems pathological to me; for he was well employed at Mannheim-Waldhof. In fact, not long afterward a mental illness developed that went hand in hand with a physical decay and in the end led him to suicide. His sad fate hit close to home, because I liked him. However, he had become so withdrawn through the illness that he no longer visited me during his employment at Schering; on the contrary, he avoided every encounter. This apparently was the harbinger of the severe illness.

Dr. Giebe, who was an assistant for several years in the lecture, suffered a similar, but also not entirely so severe a fate [as Hunsalz.] With him I carried out a somewhat lengthy investigation of the formation of the acetals from aldehyde and alcohol in the presence of hydrochloric acid. At that time, he already had heart disease but surely might yet have lived long. Unfortunately, he had the ambition to become a soldier and therefore, against the advice of the doctors, voluntarily entered the service in a hunter battalion. He was released after several months because the duties were [too] strenuous, and he died perhaps ½ year later nearly at the same time as his father. He was a splendid, ambitious, and promising young man, whose early demise saddened me greatly.

The private assistant Dr. Pinkus lasted much better. He was a helper at first, but from 1897/98 was regular assistant in the private laboratory. He carried out the experiments on the different isomeric forms of acetaldehyde-phenylhydrazone, and he was a particularly fine help in the administration of the institute. When I offered this to him, he declared to me that he had not the slightest business experience and had never kept a book [of accounts.] However, I trusted his talents, and as it happened, in all of these things, one could say, he proved to be sensible and practical. He soon safeguarded the laboratory from useless activities. Later he went to Nölting in Mühlhausen to tour the chemical dye [factory] there, and then he entered the publicly traded aniline factory in Berlin as a chemist. He later left this [firm] in order to become independent.

Dr. Hübner, born in Kreuznach, was a splendid successor to Hunsalz in 1897. He was a very deliberate, calm, and conscientious chemist. He carried out the difficult experiments on free purine and its methyl derivatives. He also provided me steno-graphic help for 3 weeks during the autumn vacation 1898 in the composition of the biography of A. W. Hofmann, for which I had undertaken the scientific part. He currently holds a respected position at the dye-works in Höchst.

In addition to the assistants, I always found room in the private laboratory for several students working on their doctoral projects. Of these I mention the later director of a sulfuric acid factory in Duisburg Dr. Haenisch, the later technical director of the cellulose factory in Waldhof Dr. Hans Clemm, the later farmer Dr. Kopisch, [and] the current professor of pharmacy in London Dr. Crossley.

Dr. Robert Pschorr and Miss Hertha von Siemens were notable special guests in the private laboratory. The first had already applied as a student in Würzburg in the year 1892 but altered his decision when he heard that I was to move to Berlin in October, because Munich and Berlin were those cities which he had to avoid during his time as a student. Instead, he went to Knorr in Jena and also earned his Ph.D. there. In the autumn 1895, however, he came to me in Berlin, and I was able to offer him a newly vacant place in the private laboratory. Following his own plans, he worked on the synthesis of phenanthrene derivatives but received much good advice from me. After 1½ years of enthusiastic and successful work, he notified me that he wanted to undertake a trip around the world with the young Dr. Meister. I advised against this on the grounds that one should undertake such long journeys for serious purposes, not just for entertainment as so-called globetrotters. However, the preparations had already been made, so he would not be deterred. Fortunately, after perhaps ⅔ year, he returned undamaged in body, spirit, and willingness to work in order to continue his chemical investigations in the private laboratory. Soon thereafter he married a young lady from Frankfurt-on-the-Main. Through the daily and close interactions in the laboratory, I quickly grew fond of him; for he is a fine, accomplished man, musical, also capable of composing poems in the Bavarian dialect for [special] occasions, and very ready to lend assistance to other people or make them happy. With the move to the new institute, I was able to offer him an assistantship, and when Harries was appointed at Kiel, he assumed the post of division head until, in the spring 1914, he came to the technical college in Charlottenburg as successor to Liebermann. Since the outbreak of the war, he has been in the field, at the moment as major in the Bavarian army, and at the jubilee celebration of the Chemical Society he received, as previous editor of the reports, at my suggestion, the title of *Geheimer Regierungsrat*.

Miss von Siemens was the first woman admitted as trainee in the chemical institute. She was warmly recommended by Anton Dohrn, the creator of the zoological station in Naples, with whom she was friends. She had already studied several fields of natural science and now wanted to educate herself in chemistry, particularly organic, in order to be able to pursue biological studies in Naples. She was already 29 years old at that time, and as she had likely had no time to engage in fundamental chemistry, I took her into the private laboratory for a winter semester, where she carried out organic preparations and several analyses and was entrusted to the special care of Dr. Hübner. She took the matter quite seriously, but naturally it was also not possible for her quickly to master so difficult a science as chemistry, particularly as I was not able to retain her in the private laboratory, due to lack of space, for longer than one semester. To express her thanks for the instruction she enjoyed, she repeatedly invited several young, unmarried assistants to the splendid villa in Charlottenburg where she and her mother lived. In this way she made the acquaintance of Dr. Harries, who married her in the autumn 1899. So, she remained permanently faithful to chemistry and later expressed her interest in our science through generous gifts, in particular to the benefit of the Kaiser Wilhelm Institute for Chemistry. She is a distinguished woman, who perhaps might have done good things, had she come to the science earlier and remained permanently.

Since then, many women have come to our institute, and during the war they have even attained the majority of the trainees. Exactly as with the young men, their performances are extraordinarily different. Among them there are perfunctory, also careless elements, who push into the lecture halls and institute more for show or for entertainment. But most, particularly among the German women, think more seriously. Among the girls who have attended the institute in the last [few] years, I have got to know 2 or 3 who were entirely equal in accomplishment with good [male] chemists. Nevertheless, I have not been able to convince myself of the advisability of women studying chemistry, because experience teaches that the majority of women who study, even the best, later marry and then typically are no longer capable of practicing their profession. As soon as this happens, the money and work devoted to the study are lost, and the same holds true for the great effort that must be expended by the docents of chemistry in the laboratories. For the time of the war and perhaps also for several years of peace it will not be possible to do without the help of women in chemistry, because the men are lacking. Also, the danger of marriage has become smaller for the moment, but as soon as normal circumstances return, my above view will again come into its own. I hope, therefore, that in future studies, women will be more restricted to those fields where female scholars can address a true need. That is the case in certain branches of medicine and above all in the teaching profession; for I expect that also among us [in Germany] the time will come where instruction in the elementary school and middle school, even for boys until they are perhaps 11–12 years old, [will be] entrusted to the woman without encountering the danger of rearing a masculine youth that is too feminine.

Toward the end of the nineteenth century, two men appeared in the series of assistants who, to my pleasure, wanted to dedicate themselves to inorganic chemistry and later also achieved the most beautiful success in this area. They were both trained in our institute. The one was a Swabian, Otto Ruff, a fully-trained apothecary, but with an *Abitur*, and fully-trained with us as a chemist. He carried out his dissertation as an organic [chemist] with Piloty, then independently found a degradation of sugars that was simpler than the older procedure of Wohl. Somewhat later I published with him a little project on the configuration of xylose and the synthetic transition of mannitol to dulcitol [galactitol]. Then he occupied himself almost exclusively with inorganic projects, for which I also felt obliged to him, in a manner of speaking, when, in October 1897, he took over a teaching position in the analytical division. He became High Assistant here 1900, and in the spring 1903 head of the division. 1½ years later he became [full] professor of inorganic chemistry at the technical college at Danzig and is currently occupied in the same capacity at the technical college of Breslau. Ruff is a talented, very diligent, and energetic chemist, who spares no pains to undertake and carry out difficult experimental projects. He also enjoys a good reputation as teacher. Once, in the interest of the institute, I derived a benefit from his inclination to fight for his convictions. In the construction of the new building, one had neglected to obtain a cost estimate for small rolling wagons for the transport of chemicals within the institute. Consequently, the firm involved delivered two wagons that were for us unusable and issued a bill that was far too high. Legal proceedings thereby ensued, which Ruff led for the institute with pleasure and comical vigor, and for which he won a sparkling victory. His joy in this dispute also brought him into conflict here and there with the contemporaries. However, for myself, I have always found him to be an understanding coworker ready for every useful activity. In recognition of his experimental research, it gave me pleasure repeatedly to provide him financial support from the Academy of Sciences or other sources.

The second inorganic [chemist] was Alfred Stock, who became the next lecture assistant in October 1898. He also performed [his doctoral work] as an organic [chemist,] which did his experimental training no damage. Stock is a child of Berlin and had already so distinguished himself at the Friedrich Werder *Gymnasium* that he was granted from there a three-year stipend for study. He was also very useful as lecture assistant and acted deliberately to find new experiments or to improve the old ones. As his inclination toward inorganic chemistry was certain, I advised him in the autumn 99 to go to Moissan in Paris, and I secured for this purpose a travel stipend for him from the culture ministry. Moissan liked him so well that he asked me in the spring 1900 to let Stock remain with him another half year. In the late summer 1900 Stock came back from Paris, filled with gratitude for Moissan, who had received him in so friendly a manner and taught him as a master of methods and apparatus common in France [but] scarcely known in Germany. For example, we thus came into possession of the so-convenient mercury trough, used by Berthelot, Moissan, and other French scholars, which, because of its size, allows very comfortable work with gases. Unfortunately, it also requires a rather expensive amount of mercury.

After his return, Stock dedicated himself exclusively to inorganic chemistry, ran through the usual leadership levels in the institute as teaching assistant and division head and in 1909 was promoted to [full] professor of inorganic chemistry at the technical college of Breslau. Several years later he was named full professor and director of the chemical institute at the University of Münster. With that, after long years, for the first time, a full professorship of chemistry was again entrusted to an inorganic [chemist,] and indeed, without the obligation to give elementary lectures in organic chemistry.

However, before Stock entered his new position, Willstätter gave up his post at the Kaiser Wilhelm Institute for Chemistry, in order to go to Munich as successor to Baeyer. Stock applied for this position and was named by the administration under the same stipulations as Willstätter. In the meantime, the war had broken out, and Stock, who was still of the age obligated to military service, immediately undertook war science projects. Also, he had to vacate the rooms in the Kaiser Wilhelm Institute, as they were taken for the gas warfare, and so he returned for the duration of the war to our institute. Stock is a very skillful experimenter, and among the younger German chemists, he might rank at the top in the construction of scientific apparatus. The proof of this was the sparkling investigation on the bonding of hydrogen with silicon and boron. In addition, he is an outstanding speaker, and I don't believe that he has an equal in any other German chemist in the inorganic experimental lecture. For me, it is therefore a great relief that he currently represents me in this lecture and thereby guards the chemical instruction in Berlin against further decline. If Stock again works as hard during the future peace as he did before, he will probably become one of the most outstanding representatives of inorganic chemistry in the world.

In the winter semester 1899/1900, the last in the old institute, Otto Diels received an assistantship. He had previously carried out his doctoral work under my direction and became Stock's successor, first as lecture assistant. The new version of the lecture book is due to him. A talented draftsman, he decorated the book with many handsome illustrations. He is the son of the classical philologist and current secretary of the Academy of Sciences and was determined early on, just like his brothers, if at all possible, to adopt the academic career. He later became teaching assistant in the organic division, and, after Pschorr left, became his successor as division head. In the meantime, he made an esteemed name for himself through the discovery and systematic investigation of carbon suboxide. During the war he became successor to Harries at Kiel, after W. Traube had declined the same appointment. Diels is a good experimenter and also quite popular as teacher. For many years he held lectures on inorganic chemistry for me. He also composed an outline of organic chemistry, very much used in Berlin. To me, he was always a dear pupil and colleague.

With the opening of the new institute in the Hessen Street, which had approximately three times as many workstations as the old building and technically much better equipped, the number of assistants naturally grew. At the same time three positions for division heads were created. Two of these went to the two leaders of the analytical division, and the third was designated for the organic chemistry. This conformed with the division of space in the institute into four equally large classrooms, of which one, equivalent to a division, was managed by me.

The first division heads were Gabriel and Harries. The third position was administered provisionally by Ruff, who received the position of Upper Assistant for this purpose and 1903 rose to division head. Among the new assistants there were the already previously mentioned R. Pschorr; further Dr. Lehmann, who accomplished so much with the new building and later entered the dye factory [of the] previous F.Bayer & Co. in Elberfeld; and finally Dr. A. Wolfes and Dr. Poppenberg. As my private assistant, Wolfes provided superb service in the projects on amino acids and polypeptides. He also took part in the experiments on Veronal, and that was likely the reason why he was recruited for the scientific laboratory by the firm E. Merck in Darmstadt after he had fulfilled his duty year of military service. In this position he proved his worth and was valued by the firm as a splendid and trusted coworker, just as it happened with me. During the war he became lieutenant in an infantry regiment, and despite his apparently not strong body, he well withstood the great exertions of the field on the western front.

Poppenberg was instructional assistant in the analytical division. He later became teacher and professor at the artillery school in Charlottenburg. He also occupied himself scientifically in the area of artillery, and during the war he made several ground-breaking inventions in his field. Unfortunately, while carrying out an analysis for nitrogen in his current position, he lost an eye through [an accident with] a drop of potassium hydroxide solution.

In the winter 1900/01 Dr. Rohmer also entered into the series of assistants and remained several years in the analytical division, where he made several small inventions, for example, improvement of the arsenic determination through distillation as arsenic trichloride. He then entered the service of the dye works in Höchst and had good success here as technical inventor. At the same time Dr. Franz, son of the proprietor of the Siechen beer house, was my lecture assistant.

Dr. Wolf von Loeben, scion of a noble officer family in Saxony, was a man of distinctive type. He had carried out his doctoral exam with Behrend in Hannover on the γ-methyl uric acid and was then, for a short time, coworker in the thermochemical investigations of Stohmann in Leipzig. He next carried out a small project with me in the uric acid group and became helper in the thermochemical division, which I had established in the new institute. It was very convenient for me that he had brought his experience from the Stohmann laboratory. The results of our joint experiments are recorded in several papers in the session reports of the Berlin Academy of Sciences. Loeben was a comfortable Saxon who enjoyed general popularity but did not like immoderate work and therefore also shied away from the chemical industry. Instead, he got the position of Assistant in a scientific research institute of the imperial treasury, which was under the leadership of Professor K. von Buchka. He died here shortly before the war, in consequence of a pyemia from a small wound. Despite his personal good nature, he was politically an outspoken jingoist and as such was at the head of the opposition in the Chemical Society, which disputed the election of Sabatier as an honorary member for political reasons. At that time, I had to oppose him rather strongly, which, however, had no influence on our friendly personal relationship.

Aside from the regular assistants in the new building, several helpers, part salaried, part unsalaried, took part in my projects in the private laboratory. Among these were the gentlemen Dr. Bethmann and Dr. Hagenbach, a son of the well-known professor of physics at the University of Basel. Later they both entered the Höchst dye works, and, through good accomplishments, Dr. Hagenbach established quite a well-respected position for himself here.

Dr. E. Frankland Armstrong, son of the professor Henry Armstrong in London, was in the same vein. He was occupied [in my laboratory] for several years after he had previously carried out an investigation in the purine group and then was granted the Dr. phil. by the Berlin faculty. He is a speculative and also experimentally well-inclined chemist, who carried out with me the somewhat difficult experiments on the synthesis of disaccharides and the preparation of acetohalogen glucose. Unfortunately, through my later observations, the existence of the α- and β-bonding designated isomers of chloroacetyl glucose has become very doubtful. In the conversion of our original preparation into α-methylglucoside, which had served as proof for the structure of the putative α-halogen body, either an error or an accidental change of configuration must have occurred.

Armstrong later wrote a handsome little book on the simple carbohydrates that was translated into German by Dr. Eugen Unna. Further, he carried out interesting experiments in London on the cleavage of the α- and β-glucosides with enzymes. However, he was [then] obliged, likely on financial grounds, because he recently married, to take a position in industry, which naturally brought a halt to his purely scientific projects.

Among the new assistants of the winter semester 1901/02, Dr. Georg Röder and Dr. Alfred Dilthey are particularly to be named. The first one carried out the syntheses of uracil, thymine, and similar compounds with me by new methods. He is a talented and also theoretically well-trained chemist, who would certainly have done handsome things, had he had the necessary perseverance. However, after several semesters he preferred to leave the laboratory and travel. From time to time he pops up in Berlin again and has interesting experiences to report. If I am not mistaken, he got to know several foreign continents and finally was employed in the laboratory of Piutty in Naples as a teaching assistant. He was chased away from here by the war, returned to Berlin and then soon became a soldier. As I understand, he soon found, through intervention of his chemical friends, a position with an *Oberkommando* of the army, where, despite his low military rank, he was happily able to make use of his chemical and technical knowledge.

Alfred Dilthey was the youngest son of my deceased sister Mathilde, very handsome and bright. After he had studied one semester in Geneva, he came in the autumn 1895 to Berlin and remained here several years. He had only the *Abitur* from an *Oberrealschule*, and the Berlin faculty was inclined in such cases to make difficulties, so on my advice he went to Hantzsch in Würzburg for his doctoral work. Then he served one year of voluntary [military] service in a lancer regiment in Düsseldorf and returned 1901 to Berlin. Here he took part in my projects on the amide structure in alkyl malonate esters and in the syntheses of alkylated barbituric acid [derivatives,] of which Veronal has become a well-known sleep aid. He had no inclination toward a scientific career, so in 1902 he made a journey to the United States of America and returned from there ¾ year later, unfortunately rather ill with malaria.

He needed a long time to recover and then settled down in Berlin in order to start [his] own business. He declined my advice, either to enter a chemical factory or to take part in the prosperous cotton-spinning mill of his father and brother in Rheydt, and he remained without success in his own business ventures. And so it came to pass, that for the talented, clever, and also diligent young man, a proper career position corresponding to his abilities and his resources remained unattainable. In this connection he became, so to say, a sacrifice to his preference for the big city. At the outbreak of the war, he went into the field. He was a deputy officer with his convoy in the summer 1915 when they were attacked by Cossacks, and he perished there. He rests in the Polish soil.

As long as he was in Berlin, he was regularly my guest on Sundays and holidays. He took pains to educate the young cousins, my sons, in all arts of masculine youth, such as playing cards, drinking wine, the history of sports, student activities, questions of dance, discourse with women, and such. The boys listened to his instruction, particularly when they were still in school, with astonishment and deep respect, and the four young people formed an excellent quartet through conviviality and harmony. Sometimes it seemed as though Alfred Dilthey belonged among my sons. As a consequence of this unholy war, of these four splendid people only one remains.

Three gentlemen were in the private laboratory at the same time as Dilthey. Despite differences in their natures, they maintained good friendships among each other and were also united in boisterous pranks from time to time. The most enterprising of them was Dr. Theodor Dörpinghaus from Elberfeld, a handsome man distinguished by bodily strength, practiced in many a sport and inclined toward an adventurous life. In chemistry he distinguished himself less through fineness of observation than through rapid, energetic attacking of experimental assignments. He carried out the first somewhat arduous hydrolysis of keratin and similar proteins. Later he went traveling and stayed, not only in America and Asia, but also fairly long in central Africa and Morocco. The result of the last journey was a bill of indictment against the administration of the Belgian Congo, which evoked some notice. Since outbreak of the war, I have heard nothing more from him. I suspect, however, that he took an active role [in the war] and emerged undamaged; for he always showed skill and had good luck in overcoming danger. In the last time of his stay in Berlin, he was, together with his friends Dr. Ernst Königs and Eduard Andreae, owner of an old sailboat on which he spent every free day of the summer and also sometimes showed off his water skills to us on the Wannsee.

Dr. Peter Bergell, the son of a Mecklenburg farmer, is a medical doctor. Before he came to me, he was an assistant at the clinic of internal medicine in Breslau. In consequence of his good [academic] preparation, he quickly found his form in our work environment. He worked on the isolation of the amino acids as derivatives of the β-naphthalenesulfonic acids. Using this reagent, we succeeded in showing evidence for the first time already 1902 that in moderate hydrolysis of silk fibers a dipeptide, glycylalanine, arises. After leaving our institute Bergell became first the medical advisor of a factory for pharmaceutical-chemical preparations in Berlin. Later he returned to pure medicine and at the moment has a large practice as specialist in metabolic disease. His friends praise his great skill in professional matters. Lately he has become a writer and has published a much-read book "The Left Landgravine" in which he deals with the problem of bigamy.

Entirely differently disposed was Dr. Hermann Leuchs, a quiet, scholarly type, distinguished by calmness, solemnity, and remaining silent. He stemmed from a manufacturing family in Nürnberg, came to Berlin at the beginning of the century, and carried out a doctoral project under my leadership on the synthesis of oxyamino acids. The most beautiful result was the artificial preparation of serine and glucosamine. The skill and care that he showed in this [project] was the inducement for me to select him as private assistant for the difficult project on polypeptides. After he had rendered superb service here for two years, he became instructional assistant in the organic division and finally, after the departure of Diels, his successor as division head. Leuchs is an exceedingly clever chemist and a very skillful experimenter. This was proved in his later investigations, particularly the analysis of the alkaloids of the *nux vomica*.

Together with Dr. Dilthey, these three young men in the private laboratory, despite assiduous and also successful work, led a cheerful life, which rose from time to time to boisterous pranks. This was particularly the case when, in the winter 1903/04, I left Berlin for several months due to sleeplessness and roamed around the Alps. This awakened a feeling of independence in the youth that was manifested first in inventions and patent applications, which afterwards, however, proved to be worthless. Along with that came all possible practical jokes. [For example,] an entire laundry basket of fresh eggs that Dr. Bergell wanted to use for a scientific experiment was secretly cooked, and thereby the original purpose [of the eggs] was spoiled. Then there was a friendly tussle among the tall, mostly very strong gentlemen, in which Dr. Bergell lost a part of his beard and Dörpinghaus nearly lost an eye [after] taking a foot to the face. From this one can infer that other things happened in the private laboratory besides pure science, particularly when my absence from Berlin removed the danger of being surprised. Fortunately, at that time there was not yet a female element among us. Otherwise, difficult complications might have set in.

Contemporary with the just-named gentlemen, two new assistants in the inorganic division had been named: Dr. Blix, a Swede, and Dr. Arthur Stähler, from Berlin. Blix had previously completed his doctorate in Berlin with moderate success, but he was praised by the leader of the division for his experimental skill. Then there was his dignified personality and his winning manner. But later difficulties with other assistants showed that he was not comprised entirely of sweetness, and he gave up the assistantship before the end of the two years. He then went to the United States of North America [*sic*] and there, as I learned indirectly, obtained an excellent position in the sugar industry.

Dr, Stähler remained in science as an inorganic [chemist.] At my suggestion he was sent by the culture minister 1906 to Th. W. Richards at the Havard [*sic*] University in Cambridge (Mass.) There he took part for about a year in the well-known determination of atomic mass. Consequently, he became the trusted helper of Richards when the latter came to Berlin in the summer 1907 as exchange professor. There followed from this collaboration several joint publications by Richards and Stähler on atomic weight. Later, Stähler continued the experiments on his own, while also being occupied with tasks in mineralogical chemistry. After the outbreak of the war, he entered the service of the War Metal Company and for several years has led an analytical laboratory in Brussels and finally in Cologne-on-the-Rhine.

Name Index

Subject Index

Printed in the United States
by Baker & Taylor Publisher Services